Risk, Language, and Power

Risk, Language, and Power

The Nanotechnology Environmental Policy Case

Jeffery T. Morris

LEXINGTON BOOKS
Lanham • Boulder • New York • Toronto • Plymouth, UK

Published by Lexington Books
A wholly owned subsidiary of The Rowman & Littlefield Publishing Group, Inc.
4501 Forbes Boulevard, Suite 200, Lanham, Maryland 20706
www.rowman.com

10 Thornburg, Plymouth PL6 7PP, United Kingdom

British Library Cataloguing in Publication Information Available

Library of Congress Cataloging-in-Publication Data

Morris, Jeffery T., 1961–
Risk, language, and power : the nanotechnology environmental policy case / Jeffery T. Morris.
p. cm.
Includes bibliographical references and index.
ISBN 978-0-7391-7054-0 (cloth : alk. paper) — ISBN 978-0-7391-7055-7 (electronic)
1. Nanotechnology—Environmental aspects—Government policy—United States. 2. Environmental policy—United States—Public opinion. 3. Deliberative democracy—United States. I. Title.
TD196.N36M67 2012
363.18'9—dc23

2011050051

™ The paper used in this publication meets the minimum requirements of American National Standard for Information Sciences—Permanence of Paper for Printed Library Materials, ANSI/NISO Z39.48-1992.

Printed in the United States of America

Contents

List of Tables vii

List of Figures ix

Acknowledgments xi

1 Introduction 1

2 Risk, Regulatory Science, and Nanotechnology 31

3 Language and Nanotechnology 59

4 Power and Its Grip on Environmental Discourse 91

5 Implications for Environmental Risk Debates 119

6 Conclusions and Proposed Path Forward 153

Appendix: Charts Describing How Statements Can Be Organized 173

Bibliography 177

Index 187

About the Author 193

List of Tables

Table 3.1 Discourse Analysis Summary 62
Table 5.1 Summary of Implications of Key Findings on Power Relations, STS Scholarship, and Targeted Policy Initiatives 127

List of Figures

Figure 1	Image of nano cubes	6
Figure 2	Nanotechnology Governance Relationships: An EPA-centric View	164
Figure 1A	Sources of Statements (as a percentage of total number of statements)	173
Figure 2A	Sources of Statements	174
Figure 3A	Statements Within Discursive Categories (as percent of total number of statements)	174
Figure 4A	Statements Within Discursive Categories, Grouped by "Sector" (as percent of that group's statements)	175

Acknowledgments

I owe a debt of gratitude to numerous people for the support, encouragement, and advice they gave me as I pursued this project. I cannot thank enough Barbara Allen for the excellent guidance and insights she provided throughout the development of this work. I wish to acknowledge Michael Hall for his longstanding friendship and interest in my academic endeavors. My parents, Bill and Nadine Morris, always believed in me and buoyed my resolve with their continual encouragement. My son, Tomas Morris, inspired me with his constant good humor, and his interest in my pursuits helped affirm that my endeavor was worthwhile. Most of all my thanks go to Tracie Greenway Morris. In this project she went far above and beyond the role of supportive spouse, providing editorial and technical advice. But more than that, she has always been unwavering in her belief in me and in the value of what I have tried to accomplish.

1

Introduction

On April 10, 2006, a story broke in the technology trade press that the use of a German aerosol glass and ceramic sealer called Magic Nano had sickened nearly one hundred people and hospitalized six of them with symptoms of respiratory illness.[1] This was the first reported incident of human injury allegedly linked to nanotechnology.[2] As it turned out, several weeks later the German Federal Institute for Risk Assessment concluded that Magic Nano did not contain nano-scale particles.[3] Yet as I sat in a hearing room of the Rayburn House Office Building five months later, listening to a panel of industry and government witnesses discuss science priorities for understanding the environmental implications of nanotechnology, I was tapping away on my Blackberry in response to emails indicating that anti-nanotechnology groups were continuing to cite the Magic Nano incident as justification for their call for a moratorium on nanotechnology until its risks are sufficiently understood to set scientifically defensible safety standards.[4] Irrespective of whether in the fall of 2006 the officials testifying before the House Committee on Science had moved beyond the early-spring Magic Nano incident, debates on nanotechnology risks were just beginning.

Fast forward four years to May 20, 2010, when the J. Craig Venter Institute announced that it had synthesized the first self-replicating synthetic bacteria cell. The synthetic genome was designed on a computer, chemically synthesized, transplanted into a yeast cell, and two days later the self-replicating *Mycoplasma mycoides* JCVI-syn1.0, containing only the Venter Institute's synthetic DNA, came into existence in a Petri dish.[5] One day after the Venter Institute's announcement, President Barack Obama directed the Presidential Commission for the Study of Bioethical Issues to make artificial DNA a top priority; a week later, Congress called Venter to Capitol Hill to testify on the implications of his institute's breakthrough.[6] In his opening remarks at the hearing, Energy and Commerce Committee Chairman Henry Waxman stated, "The promise of synthetic biology does not diminish the importance of its being conducted and applied responsibly. As is true whenever science advances, we must weigh and manage the safety, health, and environmental risks posed by this evolving science."[7] Senator Waxman's call for enhanced responsibility for managing new technologies was not new or unique to synthetic biology. Members of Congress have made many similar statements about nanotechnology risks. They also have made them repeatedly over the decades for pesticides and other chemicals. Yet the existence of tens of thousands of minimally tested chemicals currently in commerce—together with long-standing and unsettled debates over the risks of many of these chemicals—suggests that American public policy has not been effective in shaping consensus on how to discuss, let alone address, the environmental risks that accompany the introduction of new technologies into society.

I argue that the treatment of environmental risk in public policy debates in many respects has failed for industrial chemicals, is in danger of failing for nanotechnology, and most likely will fail for synthetic biology and other new technologies unless we change how we describe the impacts to people and other living things from the development and deployment of technology. However, I also argue that the nanotechnology case provides reason for optimism that risk can be given different, and better, treatment in environmental policy debates. Therefore, in this book I explore discourse around the environmental risks of nanotechnology, and through this study of nanotechnology make the case that the dominance in risk discourse of regulatory science—the use of science to identify and characterize risks for the purpose of regulation—is limiting policy debate on environmental risks, and that specific initiatives should be undertaken to broaden debate not just on nanotechnology, but generally on the risks of new technologies.

The characterization of environmental risks in the American public policy arena is too narrowly focused. The dominance of regulatory science in environmental debates reduces risk to an articulation of hazards and exposures expressed in technical, and technocratic, assessment documents. While the assessments of risk produced by regulatory science are useful, they are limited and using them as the sole or even primary bases for environmental protection discussions and actions excludes from policy debates voices that could broaden and enrich discourse around the implications of emerging technologies. Indeed, as narrowly focused as they are, regulatory science assessments are filled with value judgments and subjective interpretations of experimental observations.[8] Issues that do not get addressed in regulatory science assessments of risk include the cultural impacts of technology development, distribution of benefits and impacts across society, ethical considerations, and the effect of new technologies on existing socioeconomic disparities in society.[9]

Richer debate in my view not only would provide new decision options for policy makers, but it also would strengthen democracy by opening participation in environmental policy discussions to individuals and groups who are either not conversant in regulatory science or see other approaches and perspectives as being equally or more useful for addressing technology-induced environmental stressors. In over two decades of US federal public service, most of which has been at the US Environmental Protection Agency (EPA) working on science policy issues, I have seen little broadening of the definition of risk beyond the boundaries of regulatory science. After working six years on nanotechnology issues, for three years of which as EPA's National Program Director for Nanotechnology, I see the nanotechnology case as illustrating both the shortcomings of regulatory science-focused understanding of risk, as well as where opportunities exist for complementing regulatory science with other considerations about how risk should be considered when making decisions about the impacts of technology on humans and the natural environment.

I believe that how we as a society talk about risk, and who talks about it, will shape the larger debate (assuming there is one) about how future technological advances are addressed in environmental policy.[10] Therefore, in this work I make risk the focus, language the subject of study, and power the motive force driving and directing risk debates as I examine nanotechnology as a case to illustrate a more-generalizable dynamic between risk, language, and power. Because how we talk reflects how we think, looking at language is a good starting point for enhancing environmental policy in a manner that will help us confront the challenges society will face as we strive to increase our wisdom concerning environmental sustainability to a level on par with, or at least close to, our cleverness in manipulating matter.

Aspiring to something close to an environmentally sustainable relation with technology is not only crucial for the future of society, it is important to me personally and professionally. While writing this work, I entered what likely will be the final decade of my career as a federal policy official. Occupying a senior position in my agency's career (i.e., non-political appointee) ranks, I have had the privilege and opportunity to both view first hand and to some degree shape policy makers' discussions of the environmental implications of technology. Since I have staked out the position that current risk debates are too narrow, it seems incumbent upon me to try to do something about it. This book is a first attempt, from an academic perspective, to do so.

Deciding to explicitly locate my professional status within this academic work strikes me as something of a gamble. The danger is that I will too often substitute professional judgment for scholarship, reaching back into experience or past observations instead of drawing from the literature. However, I see the main benefit as transparency: the reader will know, up front, that I come to this work with a specific type of experience and institutional background that has shaped my approach to environmental science and policy.[11] Hopefully the benefits of this openness outweigh the dangers.

WHAT IS NANOTECHNOLOGY, AND WHY MAKE IT A CASE STUDY?

As long as humans have worked with materials, be it carving stone implements, building wooden ships, making bronze shields, or constructing steel bridges, there likely has been a desire to alter the properties of those materials to more-closely match the fabricator's vision of the finished artifact. Indeed, the perceived limitations of materials largely shape the human imagination of what is possible to achieve in the fabrication of technological artifacts. The desire by technologists to push the limits of the properties of materials provides us with an opportunity, through the nanotechnology case, to examine various aspects of the social construction of technology.

Peter Kroes reminds us that with material artifacts there is a relationship between social construction and material properties.[12] The design of an artifact typically begins with some identified need or want, and the function of the artifact is circumscribed by human ingenuity and knowledge. Likewise, the realities of physical materials—their capabilities and limitations—set the parameters for what is possible in developing a workable design for a new material and the articles from which it is fabricated. The nanotechnology case illustrates the relationship between the social desire to enhance the properties of the raw materials from which we make our artifacts, and the physical limitations of the materials themselves.

Nanotechnology is the control of matter at the atomic and molecular scales to produce new materials and applications.[13] In 1959 physicist Richard Feynman gave a famous speech to the annual meeting of the American Physical Society at Caltech, during which he talked about the possibilities of rearranging atoms to make new materials. He asked: "What would the properties of materials be if we could really arrange the atoms the way we want them? . . . I can't see exactly what would happen, but I can hardly doubt that when we have some control of the arrangement of things on a small scale we will get an enormously greater range of possible properties that substances can have, and of different things that we can do."[14] The real possibility of achieving such control began to be realized about three decades after Feynman's speech, as advances in electron microscopy and materials science led to "cluster-assembled" molecules of such substances as titanium dioxide that were early precursors of today's nanotechnology.[15] Today, particles smaller than 100 nanometers are made from carbon, clay, and silica; metals such as gold and silver; metal oxides including titanium dioxide, zinc oxide, and cerium oxide; and numerous other substances. Because these particles have properties that are different from those of larger forms of the same material, the particles are incorporated into industrial materials and consumer products to enhance or change the performance of the product. These nanoscale particles ("nanoparticles" or "nanomaterials") are used in such products as composite materials, textiles, paints, and coatings; are being tested for medical uses such as targeted drug delivery and therapeutic treatment; and are being developed for use in electronics and other technology applications.

As of late 2009, the number of products that contained nanomaterials or in some respect were nanotechnology enabled was estimated to range between 500 and 1,000.[16] It appears impossible at this time to obtain an agreed-upon estimate of the total dollar value of such products, or to what extent the addition of nanomaterials to the products contributes to that value. Annual global private- and public-sector spending on nanotechnology research and development is estimated to be between ten to fifteen billion dollars, and is reflected in steep annual increases in the number of nanotechnology patents filed (up to over 12,000 total patents in 2006). Tens of thousands of nanotechnology papers are published each year in the scientific literature.[17]

Nanoparticles are made in a variety of ways. Some are fabricated using lithographic techniques, whereby various technologies—such as electron microscopy, X-rays, and magnetic fields—are used to create patterned substrates from which uniform particles are extracted. Other techniques apply energy to graphite and metal catalysts to make single- or multi-walled carbon nanotubes. Some nanoparticles are grown as crystals, and others are created using a combination of approaches, such as growing crystals on a substrate created through nanolithography.[18] In order to observe, measure, and manipulate nanoparticles, nanotechnology at both the laboratory and industrial scales relies heavily on imaging tools such as transmission electron microscopes (TEM), scanning tunneling microscopes, and Raman spectrometers. Figure 1 is an electron microscope picture of nano-scale cubes created through a process developed by the National Institute of Standards and Technology. The highly regular shape of the cubes demonstrates the precision control that can be achieved in producing materials through nanotechnology.

Ethical concerns can be raised at the introduction of any technology. A major ethical issue for nanotechnology relates to risks and benefits. As with most if not all new technologies that preceded it in America, nanotechnology has first and foremost been ushered onto the stage of modernity with

Figure 1. Image of nano cubes

much fanfare about, and bold prediction of, tremendous benefits that will accrue to society through its introduction into our lives. Also predictably, possible risks have been acknowledged only grudgingly by nanotechnology supporters—and even then only by that subset of supporters who actually take environmental risks into consideration[19]—and are treated as so much chaff to be winnowed out of the scientific process so that the pure grains of net societal benefit can be gleaned from the technology.

This does not, from a product-improvement perspective, make nanotechnology's current benefits any less real or its future promise any less bright. The nanoscale metal oxide particles put into sunscreens do appear to enhance protection from solar radiation, perhaps resulting in reductions in skin cancer cases. Nanomaterial-enabled lighting and batteries may lead to significant energy savings. As the push continues for further miniaturization of electronic devices, the semiconductor industry is researching the use of nanomaterials to overcome the physics and heat-generation barriers that exist with current technology.[20] From iPods to pacemakers, the benefits of electronics miniaturization are highly recognized by society. Given the continued steep increase in the number of patent applications, it seems likely that the coming decades will see the commercialization of a significant number of beneficial nanotechnology applications.

One ought to ask, of course, *benefits for whom?* Since few, if any, technological benefits come without some negative impact, how will the impacts fall across society? What will the distribution of benefits look like? Advocates for environmental justice argue that typically, those people on whom technology's negative impacts most heavily fall are not the principal recipients of the benefits of technology. Finally, are the benefits of the nanotechnology-enabled products so much greater than their non-nano-enabled equivalents to warrant the possibility of introducing new and/or poorly understood risks? Failure to ask such questions leads to our being propelled down the path of deterministic, and perhaps undemocratic, technology development.

These questions are beginning to be asked, and not just by environmental non-governmental organizations (NGOs) and other usual suspects. For instance, the Organization for Economic Cooperation and Development (OECD), under joint leadership of the European Commission and the United States, in July 2009 sponsored a conference that discussed environmental applications of nanotechnology from net societal benefit and equity perspectives.[21] Also, the United States federal government agencies that comprise the National Nanotechnology Initiative (NNI) held a workshop in March 2010 that highlighted the ethical, legal, and societal implications of nanotechnology. This conference co-located such "ELSI" discussions with scientific discussions on risk assessment and risk management approaches, with the objective of fostering a fuller dialogue on risk than is typically conducted by federal agencies.[22] It remains to be seen if this is a substantive change in considering benefit-risk issues, or whether it is merely an aberration in the traditional boosterism that to date has dominated the industrialized nations' approach to new technologies.

I chose nanotechnology as the focus of this project because nanotechnology is the first significant new materials-science development since the post-World War II boom in synthetic chemical production.[23] Nanotechnology is touted, perhaps hyperbolically, as a major innovation that will spur the next industrial revolution. At present most nanotechnology products involve first-generation nanoparticles that lend functionality to traditional materials and products, such as composites, clothing, and paints and coatings. And so far the amounts produced have been relatively small and likely have presented little exposure to human and ecological systems. However, the future points to their increased use in products as well as to more-elaborately functionalized nanoparticles. In the coming decades we likely will see even more-exotic materials, both through the combination of nanotechnology with biology and cognitive science, as well as through increasingly sophisticated manipulation of cellular and genetic material. What we learn about risk language and discourse through the examination of nanotechnology can help in the design of approaches to create more-informed public debate on the impact of new technologies on the environment. The nanotechnology case also is useful because individuals and coalitions are just beginning to frame positions on the impacts of nanotechnology. The debate so far remains largely inchoate and unsettled.

In my position as EPA's National Program Director for Nanotechnology, located within EPA's Office of Research and Development, I was responsible for managing EPA's Nanomaterials Research Program, which in many respects is a traditional environmental research program in that it investigates how humans and other living things may become exposed to nanomaterials, in what concentrations, and whether and how such exposures might adversely impact humans and the natural environment. However, somewhat less traditionally, the program also strives to identify what specific properties of nanoparticles might lead to risk, to inform material-design decisions on how such properties may be addressed or eliminated before the particles make their way into products and into the environment. My role also involved being EPA's representative for the federal NNI, as well as heading the US delegation to the OECD's Working Party on Manufactured Nanomaterials.

My work with EPA involved attending and presenting at many of the numerous conferences and meetings that are held specifically on, or are related to, nanotechnology and the environment. Through this participation I interacted with most, if not all, of the key individuals who write and speak on nanotechnology either in general or specifically on environmental issues. Indeed, I know personally and have interacted professionally with all of the nine people whose statements I analyzed for this book. One of them has received funding from EPA, although I had no involvement in decisions related to such funding. I also attended the congressional hearings (and have myself both written testimony for them as well as testified as a federal witness) and federal policy meetings where nanotechnology risk issues are discussed. Therefore, I have been a participant in, as well as an observer of, the nanotechnology risk discourse analyzed in this book. I have been part of the nanotechnology risk debates.

ACTORS IN NANOTECHNOLOGY AND THE ENVIRONMENT

In 2011, it is on a rather small stage that the principal nanotechnology environmental risk actors play out their roles. The stage is small because the actors are relatively few and, so far, the stage sets are not elaborate. There are no court controversies, so the judicial branch is absent (though waiting in the wings). There have been no verified nanotechnology accidents or deaths, so the news media is quiet, if not altogether silent. While in the United States regulatory actions have been taken on specific nanomaterials, they have raised little controversy or challenge. Congressional hearings have been held, but there has not been much debate and discussions have focused more on funding and regulatory processes than on substantive issues of risk and safety. Some environmental NGOs are engaged in nanotechnology risk issues, but by and large that engagement has been within the subdued setting of scientific conferences and not in press events or sidewalk protests. Yet while the stage is small and the actors are few, the activity generated by the players is energetic and substantive. What follows is a sketch of who these players are and how they are performing their roles.

Government

In the United States, government is both a creator and a user of nanotechnology science. Approximately twenty-five federal agencies support nanotechnology research and development, and about a dozen of those play active roles related to nanotechnology environmental, health, and safety issues. These agencies coordinate their activities through the NNI's Nanotechnology Environmental and Health Implications (NEHI) Working Group, although each agency funds and directs its nanotechnology activities in accord with its respective mission and statutory mandates.[24] While some agencies, such as the National Science Foundation (NSF), support nanotechnology environmental science research solely through academic grants, most agencies, such as EPA, the National Institute for Environmental Health Sciences (NIEHS), and the National Institute for Occupational Safety and Health (NIOSH) fund academic grants as well as conduct their own scientific research. The principal users of nanotechnology regulatory science information, as regulators, are the Consumer Products Safety Commission, Department of Agriculture, EPA, Food and Drug Administration, and Occupational Safety and Health Administration.[25]

The United States federal agencies also collaborate through the OECD. In 2006, the thirty member countries of the OECD agreed to establish a Working Party on Manufactured Nanomaterials (WPMN). The OECD established the WPMN, which in addition to OECD member nations has membership from industry, environmental and labor NGOs, and some non-OECD countries, to facilitate international cooperation on issues related to the environmental, health, and safety implications of nanomaterials.[26] Within the WPMN, governments have embarked on a collaborative program to conduct exploratory research on fourteen types of nanomaterials, across fifty-nine testing endpoints. For example, Japan and the United States are leading a collaborative research effort on carbon nanomaterials, with Japan focusing on mammalian toxicology, the United States focusing on ecological toxicity and environmental fate and transport, and other nations conducting research across a number of endpoints.[27] The intent of the program is to gain basic information on the materials themselves, as well as to determine how well existing chemical test methods perform on nanoparticles. These test methods are fundamental to the OECD's Mutual Acceptance of Data program. The program, established in 1981, mandates that data generated in an OECD member country in accordance with OECD test guidelines and principles of good laboratory practice shall be accepted by other member countries for use in environmental assessment.[28] Together with the United States, the European Commission has played a leadership role within the OECD in advancing nanotechnology regulatory science.

The United States Congress, in addition to funding agencies' regulatory science activities, has held numerous hearings on nanotechnology, some of which either have touched on or have been dedicated to the potential environmental impacts of nanomaterials. In 2004, Congress created the Congressional Nanotechnology Caucus as a means to keep nanotechnology an active issue before Congress, and it has served as a forum for seminars and presentations by business, academia, and government to members and staff from both houses of Congress.[29]

Environmental NGOs

With one notable exception, environmental nongovernmental organizations have not engaged heavily in nanotechnology. That exception is the Environmental Defense Fund (EDF). EDF engaged early and frequently with government and industry on nanotechnology. In addition to participating in formal reviews of several government research and policy documents, in 2006 EDF (then known as Environmental Defense) played a role in establishing the OECD's WPMN.[30] However, the hallmark of EDF's involvement in nanotechnology was its joint development with chemical giant DuPont Company of the 2007 *Nano Risk Framework*. The *Framework*, which outlines an approach for managing potential nanotechnology risks in an iterative manner as new information becomes available, was the outcome of a two-year partnership between EDF and DuPont.[31] This partnership brought criticism from other environmental NGOs. Immediately after EDF and DuPont released the framework, the Canadian ETC Group, Greenpeace, and the Natural Resources Defense Council signed an open letter condemning the document as "fatally flawed" and a public relations stunt.[32] However, other than periodically issuing this and similar pronouncements, such as the ETC Group calling for a moratorium on nanomaterial production, with the exception of EDF environmental NGOs have contributed relatively little to the nanotechnology public policy debate or regulatory science development.

Industry

Private sector engagement in nanotechnology regulatory science activities in general has been dependent on company size. Large chemical product companies, such as DuPont and Evonik Industries AG, have been very active in the OECD, the International Organization for Standardization (ISO), and other venues where nanotechnology environmental science issues are discussed. Some of these companies, under the Business and Industry Advisory Committee to the OECD, have committed to produce scientific data under the WPMN testing program.[33] The American Chemistry Council, the major trade association for US chemical companies, also has been active in this arena. Small- and medium-sized enterprises have not been as engaged. However, the NanoBusiness Alliance, whose membership includes many of these companies, through its lobbying and other activities has been involved in discussions about the generation of environmental testing data. Despite this activity, a 2009 EPA report indicates that relatively few companies, large or small, have been forthcoming in voluntarily submitting environmental data on their products.[34]

Academia

As generators of nanotechnology regulatory science information, academic scientists are prominent in discussions of how science should be used in environmental decision making. In the United States, a large portion of academic research on nanotechnology and the environment is funded by the federal government. Major funders include NSF, NIEHS, EPA, and NIOSH.[35] Academic researchers are invited by government to critique federal research strategies, and they sit on advisory committees and peer review panels. Federal agencies in turn attempt through their grant solicitations to set expectations for what types of scientific information would be most useful, both in terms of basic research to advance the general state of the science, as well as more-applied information for regulatory decision making.[36] While less active in international venues such as OECD and ISO, which have been populated generally with government, industry, and NGO representatives, academic researchers nonetheless are often consulted by these and similar bodies on scientific issues, and the guidelines and standards issued by international organizations and national governments help frame the procedures used by academic researchers in conducting environmental studies of nanomaterials.

Each of these groups has active players in what I refer throughout this work as "the nanotechnology debates." What I mean by nanotechnology debates is the discourse that occurs in and around the institutions that shape US environmental policy. These institutions include the three branches of government, quasi-governmental institutions such as the National Academy of Sciences, academic institutions and think tanks, NGOs, and private industry. The discourse occurs in face-to-face encounters, on conference calls and video conferences, through hard-copy and electronic media, in reports and scientific literature, and in human actions such as making an electron micrograph image or inserting nanoparticles into a cell culture.

STUDY OBJECTIVE AND ANALYTICAL APPROACH

My objective for this book is to describe how within one social context—discussion of nanotechnology in the environmental policy arena—the use of language has impacted the depth and breadth of debate on risks associated with new technologies. I explore how risk is being discussed by governments, industry, public-interest groups, and other stakeholders with respect to the potential human and ecological impacts of materials that are manufactured through nanotechnology. There are two main motivations for this exploration. First, I seek through this study to advance specific recommendations for expanding environmental debates on the risks of emerging technologies beyond the scope of regulatory science in its current form. Second, as a scholar of science and technology studies (STS), I wish to promote opportunities for STS scholarship to advance the study of how particular uses of language influence environmental policy debates.

My approach to meeting the study objective was to examine language; specifically, language concerning the potential environmental risks associated with nanotechnology. My experience in the environmental policy arena suggests that the *Congressional Record* can be a good starting point. Therefore, I consider testimony from every hearing that Congress has held where nanotechnology, nanoparticles, nanomaterials, and the like have been discussed. Since my study in not limited to debates within Congress, I have looked for strong voices both inside and outside the hearing room. This led me to select nine individuals and analyze statements they have made related to nanotechnology and environmental risks. The 665 statements made by those nine individuals are the source material for this study's discourse analysis.

Nanotechnology, as a new, partially emerged technology with economic implications and an environmental science profile close to, but decidedly different from, traditional chemicals, presents new uses of language not seen previously in the public policy arena. In addition, nanotechnology represents perhaps the first singling out in the United States of a particular technology for significant, sustained federal government support by multiple federal agencies through the National Nanotechnology Initiative. This situation raises the potential for power implications not seen in other environmental policy debates, specifically in terms of the power implications of government influence over technology development in response to global economic exigencies and how this is complicated by the power that government also holds to define, articulate, characterize, and address risks. To adequately capture the language used to describe relationships between nanotechnology and environmental risks, I have evaluated various authors' interpretations of environmental discourse and drawn upon their insights to organize my study of nanotechnology-related language within the American environmental protection episteme.[37] Maarten Hajer cites Michael Billig's terms "categorization" and "particularization" as starting points for the selection of *discursive categories* to refer to ways in which discussions are framed within a discourse.[38] Agreeing that within discourse there are different simultaneous framings of language, for the nanotechnology case I have followed Hajer's practice of organizing discourse in terms of discursive categories. My three discursive categories—risk society, technological progressivism, and administrative pragmatism—draw in particular from the scholarship of Ulrich Beck, John Dryzek, Hajer, and Daniel Kleinman.[39]

Risk society is the first of my three discursive categories. Central to Beck's risk society is reconsideration of technology's benefits through an examination of the negative impacts resulting from technology. This *reflexive modernization* comes about because over time modern society has tended to become nearly as concerned with avoiding harm as with creating social benefit through technology.[40] For nanotechnology and the environment, risk society language questions whether any benefits, demonstrated or projected, from nanotechnology outweigh its potential risks. Statements within the risk society category question whether certain applications of nanotechnology ought to be developed, given uncertainty over their safety. Also included within this category is language that alludes to the distribution of benefits and risks; that is, who are the people who are to accrue nanotechnology's benefits, and who will incur most of its risks?

Technological progressivism is the second discursive category I have created to organize language within the nanotechnology risk discourse. This category comes from Kleinman's work, who contends that technological progressivism and scientism are key discourses in American society.[41] Under the technological progressivist view, people come to see technology as moving under its own power down the only path possible.[42] The result is little questioning of the purpose, utility, or value of technologies as they enter into society. Voices within this category make statements suggesting that nanotechnology advances are part of a natural progression of scientific understanding of (among other things) quantum phenomena and technological advances in microscopy, chemistry, and materials science. Language within this category tends not to question nanotechnology's value to society. Other voices within the nanotechnology risk discourse may acknowledge the benefits of nanotechnology, but technological progressivist language tends to give unqualified endorsement of nanotechnology. This does not mean that language within this category does not recognize the possibility of unintended adverse consequences. However, it describes such impacts as unfortunate outcomes of technology's inexorable forward march—outcomes to be managed (often with technology) as we identify them.

I have labeled the third discursive category *administrative pragmatism*. This is a combination of two subcategories of what Dryzek labels a *solving problems* discourse: administrative rationalism and democratic pragmatism, together with Hajer's ecological modernization. In administrative rationalism, if problems are left to experts and their expert systems to solve, then problems likely will be defined and addressed by existing disciplinary categories of expertise, by people acknowledged by the institutional hierarchy as competent and qualified. Within democratic pragmatism, centering problem formulation and solution around "civic man" through the use of informal networks means that problems will come from grassroots concerns, but with the limitations that they may not incorporate an integrated systems view and that they may be captured by special interests and subordinated to powerful stakeholders.[43] Ecological modernization brings the perspective that environmental problems can be overcome through technological fixes.[44] For nanotechnology I combined these authors' concepts of environmental discourse into one category because I believe they effectively reside together, and indeed are inseparable, within the US public policy arena. Whether in practice or as part of an idealized notion of environmental protection governance, the language of expert opinion is intertwined with, and located within, broader ideas of the public as final arbiter of societal good. Within the administrative pragmatism category, the public sanctions expert opinion only so long as such opinion serves the common good, as defined by public opinion.

I locate within these three discursive categories all of the nanotechnology risk-related statements collected for this study. Each statement has a voice that fits best into one particular category.[45] That said, in a number of cases I encounter what I call *overlap statements*, where a statement contains attributes that fit within more than one category. In each of these cases, I identify the dominant discursive category as well as secondary and in a few cases, tertiary, categories into which the statement could be placed. Such overlaps are not surprising, since irrespective of where one draws boundaries to organize language, there will be instances where statements do not fit neatly within those boundaries. Human thought is nuanced and equivocal, particularly as people grapple with such complex subjects as risks, benefits, and the location of science and technology within society.

KEY FINDINGS

In this book I present four key findings regarding the interactions of risk, language, and power in the identification and articulation of environmental risks that came out of my study of the nanotechnology case. These findings, while interesting themselves, are but a means to the end of broadening and enriching risk debates. They lay the groundwork for reaching conclusions about, and making proposals for, broadening and enriching not only nanotechnology risk discourse, but as well the broader discourse on environmental protection.

1. Scientism, manifest in regulatory science, is a story line that runs throughout the nanotechnology risk discourse.[46] In essence, this finding means that those who control the debate on environmental risks see no alternative to science as a means of articulating risks associated with technology. Scientism demonstrates itself when people find that the only way they can legitimately challenge a technology—that is, in a manner accepted by the dominant scientism view—is on some scientifically measured grounds, such as health or ecological effects.[47] Kleinman identifies three aspects to scientism: (1) facts and values are separate, (2) facts are superior to values, and (3) scientists have "superior cultural authority."[48] Scientism's partner is technological progressivism. Important to Kleinman's argument is his view that while the social world is constructed, it is also fairly stable. This means that the development of social activities, including science and technology, will tend to follow the contours of the cultural landscape that has been shaped over time.[49]

I argue within this finding that the institutions of science, industry, government agencies, and the environmental NGOs are not prepared to lead change away from these dominant cultural perspectives. Neither may it appear to be in these institutions' interest to do so. Daniel Sarewitz points out that scientism and technological progressivism shift responsibility away from elected officials, to science and engineering, to address the impacts of science and technology.[50] Unfortunately, as Sarewitz argues, this moves political decision making away from democratic institutions and into a technocracy dominated by private interests.[51] For the nanotechnology case, such movement of power away from democratic institutions leads, in the first instance, to a philosophically thin discussion of what health and well-being mean with respect to nanotechnology; and in the second, to the likelihood that any actions taken to address safety concerns will be limited in scope and will be reactive to the emergence and deployment of nanotechnology products.

2. Language describing distinctions between natural and human-created objects causes uncertainty and instability within discourse about nanotechnology risks. I find confusion in the language of those who discuss nanotechnology, be they supporters or detractors of the technology, about whether they believe nano-scale particles to have their origins in nature, in human creation, or somewhere in between, and whether or to what extent any such distinction matters with respect to risk. This uncertainty has implications for the conduct of risk debates. In our society, naturally occurring substances tend to be characterized as less risky than those made by humans.[52] However, within nanotechnology risk debates, human-engineered nanoparticles are portrayed, in particular by technological progressivists,

as more-easily controlled than are naturally occurring nano-scale particles. Instability, as I use it here, is an outcome of uncertainty. The uncertainty over whether nanoparticles are better located within nature or as products of *homo faber* has led to instability. Specifically, I mean instability in the discourse, which is manifest in language that seems to vacillate between characterizing nanoparticles as of human creation or of being natural, as well as in practices, such as identifying data needs, that seem inconsistent or unsure in their intent. Here is one example. Statements using language that I have organized under the risk society discursive category raise risk concerns based on what is perceived as manufactured nanoparticles' "differentness" from nature. However, because the risk society language is as grounded as all other nanotechnology risk language in regulatory science, the basis for expressing such concerns rests in regulatory science making an environmentally relevant distinction between natural and human-created particles. However, the scientific community does not have such data. Therefore, instability arises out of making claims through a discursive strategy (risk society language) that cannot be substantiated because of adherence to an over-arching scientism/regulatory science narrative that cannot, through data generation and analysis, resolve the natural-human distinction.

3. Overlap statements blur boundaries between discursive categories. Statements that fit within more than one discursive category can blur boundaries by adding equivocation or uncertainty. For instance, some speakers whose statements are analyzed in chapter 3 express a risk society concern about nanotechnology's impacts, but they also make technology progressivist assertions that such concerns could best be addressed by bringing science and technology to bear on understanding potential risks from nanotechnology.

In the nanotechnology debates, language that overlaps discursive categories has the effect of reinforcing the epistemological privilege of regulatory science in policy discussions. It does so by infusing scientism throughout the discourse, directing the many uncertainties associated with nanotechnology's potential environmental risks back to science as the best reducer of such uncertainty, irrespective of whether or not the discursive category contains language presenting nanotechnology in a positive or negative light. The effect this overlapping language produces is consistent with my argument that scientism, in the form of regulatory science, is a story line that runs throughout the nanotechnology risk discourse and shapes the language of the nine individuals whose statements I have analyzed for this book.

It is well understood in the discourse analysis literature that individuals' use of language is variable, so it is not surprising that we find statements about nanotechnology risk that not only cross discursive categories, but are contradicted by statements made by the same speaker in different situations. As noted by Jonathan Potter and Margaret Wetherell, the function of an individual's statement will drive its content, and the function that a statement serves for a speaker is situationally dependent. Potter and Wetherell state that "the principal tenet of discourse analysis is that function involves construction of versions, and is demonstrated by language variation."[53] This idea of construction fits well into my view of how environmental science is socially constructed, through language and practice, within discourse.

4. Influential discourse participants have a common set of skills and attributes. The nanotechnology case suggests that people who have strong voices in policy debates possess three skills and attributes: expertise in the scientific subject matter, being present at key policy venues, and having the capability to clearly translate risk issues into policy-relevant language. The nine individuals whose language I analyzed for this book have those skills. They are the types of people who repeatedly are invited to testify before Congress, sit on government advisory committees, and give press interviews. The assertion that those who successfully participate in policy discourse have this skill set is supported by the literature, as well as by my own observations from my location within the nanotechnology policy arena.[54] Through discourse analysis of the nanotechnology case, it is possible to see not only how these skillful individuals are able to shape regulatory science information into compelling policy arguments, but also how they are able to modify their messages—moving across discursive categories—depending on the objective they are seeking to achieve within a particular contextual setting.

A challenge for enhancing environmental policy debates is that those who are participating in the nanotechnology risk discourse are steeped in regulatory science and they tend to frame risks, and solutions for addressing risks, in ways associated with scientism and regulatory science.[55] However, my analysis of their statements also shows that their language can be nuanced, and this presents an opportunity for shaping nanotechnology risk discourse. If these influential individuals can be placed in situations where their skillful use of nuanced language can be developed and translated into policy-relevant proposals for advancing discourse, this will not only advance thinking on nanotechnology, but could have broader positive ramifications for how the environmental implications of technology are discussed. Knowing what skills and attributes contribute to successful participation in policy debates can also inform strategies for recruiting new participants into the environmental policy arena.

STS CONTEXT AND SOCIAL CONSTRUCTION

I approach this project from the STS perspective that science, as with all human activities, is socially constructed. By *socially constructed,* I refer to the view that science and technology are inherently social activities shaped by society's political, economic, and cultural forces, within the systems and networks through which those forces move and are maintained. David Hess offers a broad definition of social constructivism as approaches that "trace the way in which social interests, values, history, actions, structures, and so on shape, influence, or otherwise explain the content of science and technology."[56] I find this definition useful, and suggest that we also consider how science and technology in turn shape societal views on risk. That is, to paraphrase Sheila Jasanoff's concept of co-production, to consider the ways in which risks are tied to society's relation with—and construction of—science and technology.[57] Also highly useful is Thomas Gieryn's view that science is situated within culture and that its boundaries and features are shaped by society, for society's purposes, and are strongly guided by scientists' *agency*—their capacity to purposefully influence social directions.[58]

Whether agency alone explains where the borders of science are drawn is an important question for environmental risk debates. Gieryn rejects the idea that "impersonal structural forces," in addition to agency-based motivations, can explain the scientific boundary drawing that has led to science's place of epistemological privilege in modern society.[59] It still may be helpful, however, to ask whether structural forces can be both impersonal *and* social. It may depend on how one defines "structural." Could it be that humanity has acquired an evolutionary social hard wiring that explains something about our pursuit of inquiry that cannot be explained by agency alone? Perhaps our having become human has been so firmly tied to our need to be social, that as a species we now have no choice but to view our individual identities within a larger shared conceptualization of a collective humanity that has a potential for greatness beyond what is possible for the individual. This conceptualization may be structural in that it is now part and parcel of what it means to be human; it may be impersonal in that it belongs to the community, not to the individual. Yet it is derived from our inherently social humanness. If this is the case, then with respect to science I suggest Émile Durkheim was correct in his assertion that the pursuit of knowledge through scientific inquiry is one of our *représentations collectives* that bind us together as social creatures searching for means to reconcile the reality of our everyday existence with the promise of higher-order, or *sacred*, possibilities we conceptualize for ourselves as members of liberal democratic societies.[60] The agency of individual scientists, or of particular scientific communities, indeed may be a major force in boundary drawing, but perhaps also something deeper than agency drives them to engage in map making. Could we abandon the sacred and remain human? If not, is being scientific any longer a choice for humanity? Even if it were a choice, would making the negative choice mean abandoning the Enlightenment project, and what then would replace science as our conceptualization of the sacred?

My interpretation of social constructivism does not delegitimize the value of the natural sciences in advancing our understanding of the world. It is clear from the nanotechnology case that science has been highly effective in enabling technological advances in understanding and manipulating matter. Rather, social constructivism recognizes science as a human activity that, because it is done by humans, is a social endeavor pursued within the particular cultural contexts in which its practitioners do their work. Like music, poetry, dance, and other creative activities, science at its best strives—and sometimes succeeds, occasionally spectacularly so—in leveraging human curiosity to advance discovery and awareness. Susan Haack's *sensible program* gets it about right. Haack argues that good scientists strive for "respect for evidence, care and persistence in seeking it out, and good judgment in assessing its worth."[61] There is no single generalizable, logic-based deductive or inductive scientific method. Rather, science is a "loose confederation of kinds on inquiry" that are extensions of the natural human predisposition to seek knowledge.[62] In describing her program, Haack cites Percy Bridgman's view of the scientist as "doing his utmost with his mind" and Gustav Bergmann's description of science as the "long arm of common sense."[63] That the tools of science are applied in social settings—by real people with all the idiosyncrasies, needs, wants, and desires, not to mention cultural baggage, that come with being human—does not diminish the value that the scientific endeavor brings to society.

In the nanotechnology case, and perhaps with all public policy risk issues, the social construction of science occurs within a double-caged cultural superstructure: an inner framework of quantitatively ordered liberal democratic governance that is wrapped within Max Weber's hard shell of capitalism.[64] The inner framework is a set of shared cultural representations seen generally in the western liberal democracies of North America, Europe, and Oceania, built around the idea that numbers are essential to democratic order. Nikolas Rose forms a hypothesis around the term *calculated,* arguing that numbers are used to legitimize the use of power in governance—that, in fact, numbers and democracy have a "constitutive relationship" (that is, per Jasanoff, are *co-produced*) shaped through *governmentality,* a term that Michel Foucault articulated as a way of thinking that joins the institutions of government with the modalities and knowledge systems used by those institutions to exercise power.[65] James Scott adds a measurement aspect to calculation, as a technique of government to simplify, or make legible, the complexities and locally diverse characteristics of the state.[66] Theodore Porter builds into calculation the notion of mechanical objectivity, or "following the rules," as a technology of governmentality to lend power to weak institutions.[67] Finally, Yaron Ezrahi introduces science as a "cultural resource" used in democratic politics for calculated depersonalization of political power.[68] Using these definitions, I contend that all liberal democracies function through a *calculated governance*.[69] Through the nanotechnology case, I develop this concept of calculation and its impact on risk discourse.

CHAPTER ROADMAP

Following the title of this work, the next three chapters address risk, language, and power. In each chapter, using the nanotechnology case I formulate and support a specific argument related to how the social construction of science shapes the nature of policy debates on the environmental implications of technology. While each chapter has its own theme, they build upon one another in that the risk and regulatory science discussion in chapter 2 informs chapter 3's discourse analysis, which chapter 4 builds upon for its discussion of power. In the final two chapters, I synthesize these arguments into a discussion of their implications for public policy and STS scholarship, and use this discussion as a point of departure for formulating specific recommendations for advancing scholarship and enhancing environmental policy debates.

In chapter 2, "Risk, Regulatory Science, and Nanotechnology," I develop my position that regulatory science is failing to adequately inform policy discussions on the environmental implications of technology. To set up this argument using the nanotechnology case, I outline various perspectives on risk, drawing from STS scholarship and related literature. I step deliberately and in some detail through the various components of regulatory science, joining the abundant government sources on risk assessment with my own experience and observations from working in the field. The risk and regulatory science discussions, respectively, lay practical and theoretical groundwork for the presentation of my argument for nanotechnology specifically and generally for the environmental impacts of technology, in particular emerging technologies. While my main argument in this discussion is that the use of regulatory science for nanomaterials is following the same generally unsuccessful path that regulators have taken for the past four decades regarding industrial chemicals, there nevertheless are aspects of the nanotechnology case that may serve as sources of optimism that we can begin to change the way risk is discussed in policy debates, and that there is room for shifting the location of regulatory science in those debates.

Chapter 3, "Language and Nanotechnology," makes the argument that the language used in environmental policy debates reinforces the dominance and epistemological privilege of regulatory science within the nanotechnology risk discourse. I use discourse analysis to develop and support this argument. It is for this chapter specifically that I have evaluated 665 statements related to nanotechnology and risk, asking three questions: (1) what relationships are acknowledged? (2) what assumptions are made about those relationships, and how can STS scholarship help explain those assumptions? and (3) what metaphors or other rhetorical devices are employed in the statements? In addressing these questions, I draw where appropriate from the STS literature to examine the ways in which the social construction of science and technology is manifest or described in language. What is clear from the discourse analysis is the centrality of data in the nanotechnology risk debate. Irrespective of the discursive category within which a statement may be located, the ability to speak in terms of regulatory science data appears to be essential for having a voice in the debate.

Having a voice means wielding power, and chapter 4 is titled "Power and Its Grip on Environmental Discourse." Social forces exert power within discourse in ways that limit the range of voices heard in risk debates. In chapter 4, I show how biopolitical power is shaped by liberal democracy's system of calculated governance and global capitalism.[70] I argue that there are power implications to how the quality of policy debates is impacted by the way in which risk is defined by society. For risk discourse to become more inclusive, these implications must be understood, acknowledged, and addressed through specific policy actions.

While power has many dimensions, the type of power on which I focus in chapter 4 is Michel Foucault's idea of biopolitical power.[71] Giorgio Agamben defines Foucault's concept of biopolitics as "the growing inclusion of man's natural life in the mechanisms and calculations of power."[72] For liberal democratic governments, political questions are at their root about the ordering, controlling, and sustaining of human life. Increasingly since the middle of the twentieth century, science and technology have impinged on the politics of life. As stated by Nikolas Rose, modern science and technology have allowed us "to control, manage, engineer, reshape, and modulate the very vital capacities of human beings as living creatures."[73] Science, technology, politics, and human life since the Enlightenment have been inseparable and interrelated. However, they take on new, and greater, implications for the distribution and use of power in the twenty-first century.

Chapter 5 addresses the public policy implications of the arguments I make regarding risk, language, and power, pulling those three arguments together into the single contention that STS scholarship and targeted policy initiatives must be brought together in meaningful ways (in particular, in ways meaningful to policy makers) if progress is going to be made in broadening discourse on the implications of technology for environmental protection. I examine each of the four study findings' implications for power relations, STS scholarship, and targeted policy initiatives.

Because of the centrality of power within discourse, identifying the implications of the study findings for power relations is an important prerequisite to proposing specific tools to alter such relations within discourse. The main implications of the study findings for power relations are: (1) the dominance of regulatory science places power mostly in the hands of government to define risks; (2) uncertainty over nanomaterials' place in the natural-human created/existing-new continuum destabilizes discourse over what regulatory science evidence will be required to determine the risk of nanomaterials; (3) unsettled discourse on where nanoparticles reside along the natural/human-created spectrum will cause fluctuations in power dynamics over who defines risks; (4) overlap statements demonstrate how the scientism narrative draws power away from specific discursive strategies and drives discourse toward a narrow articulation of risk; and (5) power within discourse will tend to be wielded most effectively by those with specific skills and attributes, perhaps most importantly their facility for articulating issues within a regulatory science framework.

For STS scholarship, important implications drawn from this book's findings are: (1) we need to understand how the regulatory science and technologist communities draw boundaries between their authority to define and address risks, and the claims that technologists make of the ability of new technologies to increase control over natural phenomena; (2) discourse analysis is a useful tool for exploring such boundaries; (3) there is room for generating greater understanding of the causes of instability in discourse; (4) the language used to describe environmental stressors may help reveal power relations within debates on emerging technologies; (5) opportunities exist for combining discourse analysis and ethnography to deeply analyze the cultural implications of the language used by participants in policy debates;[74] and (6) STS scholarship opportunities exist to apply ethnography to understand how the use of symbols influences social behavior.

In the area of targeted policy initiatives, five implications of the key findings are: (1) there is a lack of institutional capacity to meet regulatory science data needs, and the lack of such capacity could lead to diminished public trust in institutions; (2) regulatory science complicates the dilemma of wanting to anticipate risks but needing data to define risks, and inhibits resolving this dilemma by narrowing policy debate; (3) uncertainty over how much of nanotechnology is natural and how much is technology will shape how benefits and costs/risks are characterized in environmental policies and decisions; (4) unsettled thinking and positions on nanotechnology risks provide an opportunity for new voices to be received and considered in policy debates; and (5) understanding the skills and attributes of influential participants in the nanotechnology risk debates is instructive for crafting and implementing policy initiatives. The identification of these five implications lays the table for the final chapter's closing arguments concerning the problems with environmental risk discourse, and what to do about them.

Because positions on nanomaterial risk are still inchoate, and the application of power to advance interests does not yet appear to be well organized into coherent political strategies, there is opportunity to initiate actions that move the policy debate on nanotechnology safety in a direction of greater, more-inclusive democratic involvement. STS scholarship has important value to add. To that end, I offer in chapter 6 two summary conclusions for discourse analysis as an STS tool: (1) enhanced studies of the language of public policy can advance both STS as an interdisciplinary endeavor and discourse analysis as a tool for social scientists, and (2) there is much opportunity in STS to build off existing scholarship to explore how environmental discourse is characterized. These conclusions point to potential convergences between STS scholars and policy makers, and lead me to propose the following targeted policy initiatives: (1) hold regulatory science-focused meetings that join STS scholars with policy makers to discuss a risk-analysis path forward for one or more specific nanomaterial applications; (2) pilot real-time technology assessment within federal research and regulatory agencies; and (3) use international organizations, such as the OECD, as a venue for bringing nanomaterial testing information into case studies to locate risk information within the larger context of net societal benefit. By putting STS scholarship to work in the service of public policy, these proposals seek to use nanotechnology as a vehicle for expanding thinking generally about how to make the social construction of science present in policy debates, thereby providing greater opportunity for expanded discussions about the risks and benefits of technology. While these policy initiatives are specific and valuable in and of themselves, my larger objective in proposing them is to argue that it is specific actions, rather than general assertions or broad recommendations, that will make a difference in changing environmental discourse. I make clear as I conclude chapter 6 that emerging technologies debates should be broadened beyond risk, to focus more on language about enriching the human and environmental condition.

IN SUMMARY

Evaluating language on the environmental impacts of emerging technologies is particularly timely, as statistics on the number of scientific papers and patents in new fields such as nanotechnology and synthetic biology suggest that we are at the early stages of a wave of new products and applications with both unique properties and the potential for novel impacts on the environment.[75] As with the introduction of all new environmental stressors into liberal democratic society, these new technologies have the potential to generate substantial political debate, and will have implications for the location and use of power in decisions related to human or other biological life. Liberal democratic societies may be hard wired to be calculating. One might also imagine that were the door of capitalism's iron cage ever opened, we might not willingly venture out of it. Nevertheless, I am optimistic that there are ample opportunities to enhance the vitality of, and public engagement in, policymaking activities that directly impact the health and well-being of citizens. This book sets out to uncover how the revealing, understanding, and use of language and practice within discourse can advance such opportunities. The nanotechnology case is particularly useful in identifying what is going wrong with environmental risk debates as they relate to technology's impacts, and what can be done to get policy making in this area back on the right course. The following chapters argue that risk, language, and power are central themes that must be explored to enhance environmental discourse, and that approaching public policy from an STS perspective provides opportunities for achieving such enhancement.

NOTES

1. Kevin Bullis, "'Nano' Safety Recall: A product touted as 'nano' has hospitalized six German consumers, prompting more warnings over the dangers of nanomaterials." *MIT Technology Review*, online version, 10 April 2006, http://www.technology review.com/biomedicine/16681/p.1.
2. Nanotechnology, as described by the US government, is "the understanding and control of matter at dimensions between approximately 1 and 100 nanometers, where unique phenomena enable novel applications." National Nanotechnology Coordination Office, www.nano.gov.
3. Federal Institute for Risk Assessment, "Nano particles were not the cause of health problems triggered by sealing sprays!" BfR press release, 26 May 2006, http://www.bfr.bund.de/cd/7842. Also see George Elvin, "No nano in Magic Nano," NanotechBuzz, 27 May 2006, http://www.nanotechbuzz.com/50226711/no_nano_in_magic_nano.php.
4. ETC Group, "Nanotech Product Recall Underscores Need for Nanotech Moratorium: Is the Magic Gone?" ETC News Release, 7 April 2006,http://www.etcgroup .org/en/node/14.
5. J. Craig Venter Institute, "First Self-Replicating, Synthetic Bacterial Cell Constructed by J. Craig Venter Institute Researchers," JCVI Press Release, 20 May 2010, http://www.jcvi.org/cms/press/press-releases/full-text/article/first-self-replicating-synthetic-bacterial-cell-constructed-by-j-craig-venter-institute-researcher/.
6. David Brown, "Scientists Create Cell Based on Man-Made Genetic Instructions," *Washington Post*, 21 May 2010, www.washingtonpost.com; US Congress, Committee on Energy and Commerce, "Hearing on Developments in Synthetic

Genomics and Implications for Health and Energy," 111th Congress, Second Session, 27 May 2010.

7. US Congress, Senate, Committee on Energy and Commerce, "Hearing on Developments in Synthetic Genomics and Implications for Health and Energy," 111th Congress, Second Session, 27 May 2010.

8. Z. Meghani and J. Kuzma. "Addressing the Regulatory Agency-Industry Revolving Door Problem for Emerging Technologies: Re-Conceptualizing Objectivity," *Journal of Agricultural and Environmental Ethics* 23(6): (2010).

9. Through National Nanotechnology Initiative funding, the National Science Foundation supports individual investigator research into these topics, as well as Centers for Nanotechnology in Society at Arizona State University and the University of California at Santa Barbara. Some of this research is in the Science and Technology Studies (STS) area. However, my impression is that the questions addressed in this research are less about the ethical, legal, and societal implications (ELSI) of nanotechnology, than they are about using nanotechnology as a platform or case study to investigate broader STS or ELSI issues.

10. As noted by Kuzma and Meghani, fundamental to democracy is the idea that the broad public, and not just interest groups such as NGOs and industry, shall be the decider on the treatment of normative issues in societal discourse. J. Kuzma and Z. Meghani, "The Public Option: A possible change in the US risk-based decision making for emerging technological products: Compromised or enhanced objectivity?" *EMBO Reports* 10: 1–6 (2009).

11. I also wish to make clear that although I am an EPA employee, I am also a science and technology studies scholar, and the views I express in this book are mine as an STS scholar and do not necessarily reflect the perspectives or policies of EPA or the US government.

12. Peter Kroes and Anthonie Meijers, editors, "The Empirical Turn in the Philosophy of Technology," in Carl Mitcham, editor, *Research in Philosophy and Technology, Volume 20* (Amsterdam: Elsevier Science Ltd., 2000), 29.

13. United States National Nanotechnology Initiative, www.nano.gov.

14. This speech can be found in many locations. An online version is located at http://www.zyvex.com/nanotech/feynman.html. It was first published in the California Institute of Science's *Engineering and Science,* vol. 23:5, February 1960, 22–36.

15. Ivan Amato, "Making the Right Stuff," *Science News,* vol. 136, no. 7, 12 August 1989, 108–110.

16. The Project on Emerging Nanotechnologies, "Nanotech-enabled Consumer Products Top the 1,000 Mark," http://www.nanotechproject.org/news/archive/8277/. Over half of the products fall under the category of "health and fitness" (clothing, cosmetics, filtration, personal care, sporting goods, and sunscreen). The most-used material is silver, followed by carbon. While this article claims that the number of products has reached 1,000, the author's conversations with individuals familiar with the PEN database suggest that some products listed there are in fact no longer being marketed.

17. Nanowerk, "Ten Things You Should Know About Nanotechnology," http://www.nanowerk.com/nanotechnology/ten_things_you_should_know_6.html.

18. A useful source for an overview of nanoparticle production techniques is National Science, Engineering, and Technology Subcommittee, "Materials By Design:

Report of the National Nanotechnology Workshop, June 11–13, 2003," http://www.nano.gov/NNI_Materials_by_Design.pdf.

19. Some of the nine individuals whose statements are the subject of this study are representative of nanotechnology supporters who have acknowledged the possibility of environmental risks.

20. Linda Rae, "Nanotechnology in the Semiconductor Industry," Keithley Instruments, Inc., September 2006, 1. www.keithley.com.

21. Organization for Economic Cooperation and Development, "OECD Conference on Potential Environmental Benefits of Nanotechnology: Fostering Safe Innovation-Led Growth," 15–17 July, 2009, http://www.oecd.org/document/40/0,3343,en_2649_37417_42323688_1_1_1_1,00.html.

22. National Nanotechnology Initiative, "Capstone Meeting: Risk Management Methods & Ethical, Legal, and Societal Implications of Nanotechnology," 30–31 March 2010, http://www.nano.gov/html/about/symposia.html.

23. Biotechnology is also a significant new development, but the modification of living organisms raises a different set of issues than does creating new materials through nanotechnology. That said, it is likely that biotechnology and nanotechnology soon will find points of convergence. That possibility, and the issues it would raise, are certainly relevant to the issues addressed in this study.

24. Executive Office of the President, National Science and Technology Council, Subcommittee on Nanoscale Science, Engineering, and Technology, *Strategy for Nanotechnology-Related Environmental, Health, and Safety Research*, February 2008, 49, www.nano.gov/NNI_EHS_Research_Strategy.pdf.

25. Executive Office of the President, National Science and Technology Council, Subcommittee on Nanoscale Science, Engineering, and Technology, *Strategy for Nanotechnology-Related Environmental, Health, and Safety Research*, February 2008, 49, www.nano.gov/NNI_EHS_Research_Strategy.pdf.

26. OECD, "Safety of Manufactured Nanomaterials," http://www.oecd.org/topic/0,3373,en_2649_37015404_1_1_1_1_37465,00.html.

27. OECD, "Sponsorship Programme for the Testing of Manufactured Nanomaterials," http://www.oecd.org/document/47/0,3343,en_2649_37015404_41197295_1_1_1_1,00.html.

28. OECD, "Mutual Acceptance of Data (MAD)," http://www.oecd.org/document/41/0,3343,en_2649_34365_1890473_1_1_1_1,00.html.

29. National Nanotechnology Coordination Office, "Sen. Allen Announces Congressional Nanotechnology Caucus," NNCO press release, April 2004, http://www.nano.gov/html/news/releases/NNIConf04.html.

30. OECD, "Report of the OECD Workshop on the Safety of Manufactured Nanomaterials: Building Co-operation, Co-ordination and Communication," OECD document number ENV/JM/MONO(2006)19, 28 April 2006, http://www.olis.oecd.org/olis/2006doc.nsf/LinkTo/NT00000F92/$FILE/JT03208175.PDF.

31. Environmental Defense and DuPont, *Nano Risk Framework*, www.nanoriskframework.com.

32. ETC Group, www.etcgroup.org/upload/publication/pdf_file/610.

33. OECD, "Sponsorship Programme for the Testing of Manufactured Nanomaterials," http://www.oecd.org/dataoecd/53/41/43532644.pdf.

34. US EPA, *Nanoscale Materials Stewardship Program, Interim Report,* January 2009, 9, 33–35. http://www.epa.gov/oppt/nano/nmsp-interim-report-final.pdf.

35. Executive Office of the President, *Strategy for Nanotechnology-Related Environmental, Health, and Safety Research,* 55–88.

36. Requests for Applications from EPA's Science to Achieve Results Program are illustrative of the types of academic grant solicitations that regulatory agencies issue that attempt to balance academic freedom and creativity with the agency's mission-driven needs. The following site contains EPA's nanotechnology grant solicitations: http://www.epa.gov/ncer/nano/solicitation/index.html.

37. I use the term *episteme* in the way Foucault uses it: as the playing out of different histories simultaneously within different discourses. Foucault states that the episteme is "a space of *dispersion,* it is an *open and doubtless indefinitely describable field of relationships. . . .* The episteme is not a slice of history common to all the sciences: it is *a simultaneous play of specific remanences.*" Michel Foucault, "Politics and the Study of Discourse," in Graham Burchell, Colin Gordon, and Peter Miller, editors, *The Foucault Effect: Studies in Governmentality* (Chicago: The University of Chicago Press, 1991), 55. Therefore the American environmental protection episteme includes multiple discourses—with their own "slice of history"—around such ideas as deep ecology, sustainability, and for my study, risk.

38. Maarten A. Hajer, *The Politics of Environmental Discourse: Ecological Modernization and the Policy Process* (Oxford: Oxford University Press, 1995), 54.

39. Ulrich Beck, *Risk Society: Towards a New Modernity,* London: Sage Publications, 1992. Daniel Lee Kleinman, *Science and Technology in Society: From Biotechnology to the Internet* (Malden, MA: Blackwell Publishing, 2005). J. S. Dryzek, *The Politics of the Earth: Environmental Discourses* (New York: Oxford University Press, 2005). Dryzek follows Foucault in seeing power as a thread running throughout discourses. However, whereas the Foucauldian approach recognizes a single dominant discourse at any given historical period, Dryzek believes that environmental discourses run concurrently and in some cases interact with one another.

40. Beck, *Risk Society,* 13.

41. Kleinman, *Science and Technology in Society,* 4.

42. Kleinman, *Science and Technology in Society,* 4.

43. Dryzek, 113 and 117.

44. Hajer, 32.

45. It is a common practice in content analysis to have two people code the same statements to compare how the statements are categorized. However, given my position of being located within the policy discourse, I felt that adding another person to enable "double-checking" could be viewed as an attempt to gloss over what I admit to be a very subjective exercise conducted by someone (myself) who is an active participant in the subject of my own analysis. Therefore, I alone placed the statements within the three discursive categories and readily acknowledge that such placement is colored by my own position within the nanotechnology risk discourse.

46. Hajer defines a *story line* as "a generative sort of narrative that allows actors to draw upon various discursive categories to give meaning to specific physical or social phenomena." Hajer, 56. For instance, within environmental discourse one can follow the narrative—i.e., the story line—that science can identify and solve

risk issues, without having to adhere rigidly to a single discursive category or even understand all the categories falling within the discourse.

47. Kleinman, *Science and Technology in Society*, 4.

48. Kleinman, *Science and Technology in Society*, 123.

49. Kleinman, *Science and Technology in Society*, 13–14.

50. Daniel Sarewitz, *Frontiers of Illusion: Science, Technology, and the Politics of Progress* (Philadelphia: Temple University Press, 1996), 151.

51. Daniel Sarewitz, *Frontiers of Illusion: Science, Technology, and the Politics of Progress* (Philadelphia: Temple University Press, 1996), 160.

52. Cass R. Sunstein, *Risk and Reason: Safety, Laws, and the Environment* (Cambridge: Cambridge University Press, 2002), 36. Sunstein points out that while lay people tend to believe that naturally occurring chemicals are not as harmful as those synthesized by humans, toxicologists and other experts overwhelmingly disagree. Sunstein's discussion is based on Paul Slovic, *The Perception of Risk* (London: Earthscan, 2000), 285–298.

53. Jonathan Potter and Margaret Wetherell, "Unfolding Discourse Analysis," in Margaret Wetherell, Stephanie Taylor, and Simeon J. Yates, editors, *Discourse Theory and Practice: A Reader* (London: Sage Publications, 2005), 199.

54. Karen Litfin, *Ozone Discourses: Science and Politics in Global Environmental Cooperation* (New York: Columbia University Press, 1994). Litfin characterizes as "knowledge brokers" people who can effectively translate scientific information into language that is understandable and useful to policy makers. Steven Shapin, *The Scientific Life: A Moral History of a Late Modern Vocation* (Chicago: The University of Chicago Press, 2008). Shapin describes the attributes of scientists who have been able to effectively engage business and government interests in order to move scientific discovery from the laboratory into society.

55. This assertion is consistent with Sheila Jasanoff's examination of the role of federal advisory committees in shaping environmental policy making. See Sheila Jasanoff, *The Fifth Branch: Science Advisors as Policymakers* (Cambridge: Harvard University Press, 1990).

56. David J. Hess, "A Brief Dictionary of Science and Technology Studies." I note that this is different from Hess's definition in his 1997 work, *Science Studies: An Advanced Introduction* (New York: New York University Press, 1997), in which he refers to social constructivism as "an exogenous, independent variable that shapes or causes some aspects of the content of science and technology" (82). I cite this earlier definition in order to note that the more-recent definition is, in my view, preferable because it accounts for a more-reflexive relationship between science/technology and society. This reflexivity is important to discussions of risk as a social construct.

57. Sheila Jasanoff, editor, *States of Knowledge: The co-production of science and social order* (London: Routledge, 2004), 2.

58. Thomas F. Gieryn, *Cultural Boundaries of Science: Credibility on the Line* (Chicago: The University of Chicago Press, 1999), 10.

59. Thomas F. Gieryn, *Cultural Boundaries of Science: Credibility on the Line* (Chicago: The University of Chicago Press, 1999), 15.

60. Émile Durkheim, *The Elementary Forms of Religious Life* [1912], translation and introduction by Karen E. Fields (New York: The Free Press, 1995), xviii. A number of STS-oriented historical accounts also describe the development of sci-

ence both in terms of social construction as well as a cultural aspiration to extend the powers of human inquiry. Two examples are Steven Shapin and Simon Schaffer, *Leviathan and the Air Pump: Hobbes, Boyle, and the Experimental Life* (Princeton: Princeton University Press, 1985); and Lisa Jardine, *Ingenious Pursuits: Building the Scientific Revolution* (New York, NY: Nan A. Talese/Doubleday, 1999).

61. Susan Haack, *Defending Science—Within Reason: Between Scientism and Cynicism* (Amherst, NY: Prometheus Books, 2007), 167.

62. Susan Haack, *Defending Science—Within Reason: Between Scientism and Cynicism* (Amherst, NY: Prometheus Books, 2007), iv.

63. Susan Haack, *Defending Science—Within Reason: Between Scientism and Cynicism* (Amherst, NY: Prometheus Books, 2007), 95.

64. For a discussion of the tie between liberal democratic governance and quantification, see Nikolas Rose, "Governing by Numbers: Figuring Out Democracy," *Accounting, Organization, and Society*, vol. 16, no. 7, 673–692, 1991. Max Weber, *The Protestant Ethic and the Spirit of Capitalism,* translation by Peter Baehr and Gordon C. Wells (New York: Penguin Books, 2002), Baehr and Wells note that while it has been common to cite Weber as having referred to capitalism as an "iron cage," they contend that a more-accurate translation from the German would be "hard shell as of steel." However, iron cage proved to be a catchy term and thus became embedded in scholarly writing.

65. Rose, "Governing by Numbers," 675. Also see Peter Miller and Nikolas Rose, *Governing the Present* (Malden, MA: Polity Press, 2008), Miller and Rose break governmentality into two aspects, rationalities/programs and technologies, describing the former as "a way of representing and knowing a phenomenon" and the latter as "a way of acting on it so as to transform it."

66. James Scott, *Seeing Like a State: How Certain Schemes to Improve the Human Condition Have Failed* (New Haven: Yale University Press, 1998), 25.

67. Theodore M. Porter, *Trust in Numbers* (Princeton: Princeton University Press, 1995), 4.

68. Yaron Ezrahi, *The Descent of Icarus: Science and the Transformation of Contemporary Democracy* (Cambridge: Harvard University Press, 1990), 10, 33.

69. A term introduced to me by Saul Halfon.

70. While the political science literature does not have one agreed-upon definition of *liberal democracy,* it typically refers to representative government in which basic rights of citizens are articulated in a constitution or like document. For a discussion of where liberal democracy falls along a spectrum of more or less public participation in democracy, and the implications for discourse, see John S. Dryzek, *Discursive Democracy: Politics, Policy, and Political Science* (Cambridge: Cambridge University Press, 1990).

71. Michel Foucault, *History of Sexuality, Volume 1: An Introduction.* Translated by Robert Hurley (New York: Random House, 1978), 143.

72. Giorgio Agamben, *Homo Sacer: Sovereign Power and Bare Life* (Stanford: Stanford University Press, 1995), 119.

73. Nikolas Rose, *The Politics of Life Itself: Biomedicine, Power, and Subjectivity in the Twenty-First Century* (Princeton: Princeton University Press, 2007), 3.

74. Litfin, 37.

75. Two useful web sites for sources of industry data are www.nanowerk.com and www.nanotechproject.org.

2

Risk, Regulatory Science, and Nanotechnology

To be successful, policy debates must be well informed. Congress holds hundreds of hearings each session, probing witnesses for information related to pending legislation or other issues of national importance. On any given day in conference rooms across Washington, DC, advisory committees are convened to provide government officials with perspectives on topics ranging from securities regulation to disease prevention. Each year think tanks, academic institutions, and other organizations hold thousands of seminars and issue hundreds of reports aimed at government policy makers. Government agencies commission or conduct research to inform their policies and regulations. Information is the coin of the policy realm.

Nevertheless, policy debates on environmental risks are poorly informed. This does not mean that there is a lack of information—there are data, models, assessments, and reports in abundance. For instance, the EPA web site www.epa.gov/risk contains a tremendous amount of information on chemicals and estimates of environmental impact. As an example, searching on hexavalent chromium in EPA's Integrated Risk Information System (IRIS) brings up not only EPA-generated health hazard assessment values and the data analyses that underlie those assessments, but a long list of reports on chromium VI that go back for decades.[1] To varying degrees, similar information is available for most of the chemical substances in IRIS. Yet this type of information is inadequate for at least three reasons.

First, it is inadequate because the approximately 550 substances in IRIS represent but a fraction of the more than 83,000 chemicals on the US chemicals inventory required under the Toxic Substances Control Act (TSCA), to which hundreds of new chemicals are added every year. About 62,000 of those chemicals were in existence when EPA began regulating chemicals in 1979 and were grandfathered into TSCA, meaning that companies were not required to provide EPA with data on the chemicals' toxicity.[2] Second, the information is inadequate because even where data are available, those data do not relate effects or exposure information on an individual chemical to the wider context within which living things exist. It would be tremendously difficult, if not impossible, to know how the effects of one chemical interact with the entire mix of things to which we are all exposed, such as other chemicals, particles, UV light, and noise. There is variability within populations, and some individuals of a species are more susceptible to pollutants generally, or to specific pollutants, than are others of the same species. Also, some species are more heavily impacted by pollutants than are other species. Third, our information on the environmental risks of chemicals is inadequate because the entire paradigm for considering environmental risks is not only too narrowly focused on quantifiable expressions of risks in terms of effects and exposure, but the paradigm has failed to advance policy making on the risks from chemicals and other by-products of technology. It is on this third inadequacy that this chapter focuses: regulatory science as a faulty, failed paradigm for adequately characterizing, discussing, and addressing environmental risks.

I develop the argument first by illustrating how the scholarly literature on risk relates to the current environmental risk paradigm: how in some respects it reinforces the current paradigm, and in other cases the literature points to opportunities to either fix what is not working with the current approach or to supplant it with something else. After this review of the risk literature, I evaluate regulatory science's role in defining and assessing environmental risks. I conclude the chapter by bringing risk and regulatory science together with nanotechnology, to demonstrate through this particular case both the pitfalls and opportunities that face policy makers in evaluating the impacts of environmental stressors induced by new technologies.

RISK

In one of its more-technocratic applications, *risk* is defined as an integration of exposure and effects information to make statements about impact.[3] Noting this definition is important, as its use is central in political and public debates on risk. That said, I find the following definition by Ulrich Beck to be more useful for a broad examination of risk in society: [Risk is] "a systematic way of dealing with hazards and insecurities induced and introduced by modernization itself."[4]

The concept of risk is integral to human decision making in modernity: it is how we determine whether an activity, situation, or substance is safe, and how we make trade-offs between choosing one course of action over another. When governments address safety issues, risk becomes politicized; through their politicization, risks become instruments of power. Beck's *risk society* draws sociology, science, and technology together under the umbrella of risk. Beck argues that modernity's scientific and technological advances have resulted in a society where harmful by-products of such advances rival in social prominence the benefits derived from science and technology—we have come to realize that creating goods is nearly always accompanied by the inadvertent creation of "bads." These concerns have led to a reflexive modernization, whereby the benefits attained through modernization are opened to scrutiny and the logic of wealth creation through industrialization is questioned.[5] Modernization, once taken for granted, becomes a subject of critique. Society's treatment of agricultural pesticides is a contemporary example of reflexive modernization. During the first half of the twentieth century, the development of synthetic chemicals to eliminate pests was seen as a tremendous boon to crop yields and therefore to food availability and affordability. There was little questioning of the value of these new agricultural technologies. It has only been through decades-long discourse on the possible risks of pesticides that society has looked reflexively back at this particular aspect of modernization, identified risks, and responded with such actions as enhanced regulation and consumer demand for organically grown food.

Because risks are created "within knowledge," Beck asserts that they are socially constructed and defined.[6] In other words, as the pesticides example shows, we know of environmental risks because they have been made explicit through the generation of knowledge through science. Science is recognized as a human activity, capable of testing hypotheses but not of revealing truths.[7] In this respect, the risk society aligns with the social constructionist view of science and technology in society. In the risk society, risks created by science and technology are both defined by science and given back to science for assessment and mitigation. Because risks are invisible—meaning that by definition they are manifest in the future (transforming from risks to *events* once the negative impact occurs)—and because they are defined and constructed by society, risks lend themselves to use in the exercise of political power.[8] For Beck, environmental risks are underpinned by "a moral and social deep structure that results from the violation of survival norms."[9] Environmental risks involve the imposition of power not only directly over human life, but also over all entities and environmental processes that constitute and sustain life on earth.

In this Beck sees the potential for political reordering of a society that faces the catastrophic potential of the threats that emerge from risks generated by modernity.[10] Part of this reordering would entail a de-centering of science from its current place of epistemological privilege, as society becomes aware that science is incapable of adequately addressing the risks that science itself, together with technology, has created.[11] Science's approach to addressing risks—or, more importantly, its failure to address risks—increasingly is seen as irrational and contrary to society's well-being. Such social reordering will result in conflict, as those affected by risk confront those who create and profit from the risk-laden by-products of science and technology.[12] In Beck's vision for an advanced modernity, risks are placed out in the open for societal interpretation and discussion, and are not partitioned off as the sole purview of technocrats.[13]

Beck's vision may be both obstructed and revealed in the nanotechnology case. On the one hand, the technical complexity of nanotechnology, in particular the high sophistication of the instruments to create, detect, and see nanoparticles, limits access to the technology to an elite few individuals. This creates difficulty in making the discourse about the technology accessible to society outside science. On the other hand, the very newness of the technology at a point in history when modern society is highly sensitized to new technologies—indeed, the public's voracious appetite for the latest electronic gadget suggests that our relation to new technology goes beyond sensitization to craving—may present an opportunity to move nanotechnology further out into the open public space Beck seeks.

Mary Douglas and Aaron Wildavsky advocate a cultural theory of risk perception, whereby the entire social environment of a society determines what it will perceive as risks, and how important it will perceive those risks to be.[14] They argue that modern societies identify more risks than ever before, yet the world is safer than it ever has been. The authors believe that this perception is common across most modern societies because those societies all are grounded in the empirical and evidential ethos of the Enlightenment (i.e., are *modern*), as manifest today in both science and the law.[15] Such societies therefore accept as appropriate for policy formulation and decision making the use of formalized, quantitative risk assessments. This, Douglas and Wildavsky claim, impoverishes society's view of risk by separating out moral and values considerations. In modern society, this separation is at the core of differences in risk perception between experts and lay people.[16]

Beck and Douglas and Wildavsky are consistent regarding the impact of science's capture of the risk issue. These authors agree that discourse on risk will remain restricted so long as modern society allows risk to remain the sole domain of science. However, Douglas and Wildavsky bring to the discussion a cultural-political dimension that complicates the possibility of realizing Beck's vision. That dimension is what the authors call "knowledge as the changing product of social activity."[17] They point out that the manner in which we have constructed society determines what risks we identify as important and how we articulate the nature of those risks.[18] We have arranged our society around a capitalist and consumerist model for which science and technology are dual engines of growth. We also are a liberal democratic society in which science and the law play instrumental roles in governance.[19] These cultural factors may serve as powerful brakes on efforts to expand discourse on the meaning of risk.

Douglas and Wildavsky framed their argument decades before the emergence of nanotechnology, but it remains highly relevant today. The US federal government has invested billions of dollars in nanotechnology with the expectation of economic pay-offs through new-product development, and thereby has had a direct hand in tying nanotechnology to consumerism. At the same time, the government is legally bound and culturally expected to address nanotechnology's potential risks. Americans have come to expect technology to produce new products, but they expect them to be safe. The political and cultural location of risk therefore is an important consideration in evaluating environmental debates about nanotechnology.

Current attempts to fit the understanding of nanotechnology's environmental risks into the existing chemical risk framework suggest confirmation of Anthony Giddens's observation that trust and risk are imbricated in societal perceptions of what risks are worth worrying about and how we estimate their relative importance.[20] Giddens uses trust as a measure of belief in the reliability of knowledge to predict the probability and impact of potential outcomes.[21] Central to the creation of trust are expert systems that exist outside time and space, since they are embedded in our social structure rather than being situated in any particular location or at the hands of any particular individual. Just as with industrial chemicals, nanomaterials currently are being run through the elaborate risk analysis expert system that has been built over several decades and is relied upon by regulators, industry, and NGOs to produce authoritative conclusions on chemical risks. If the assumption by these groups is that there exists public trust in the risk analysis expert system to protect society from chemical risk, then logically they would expect the public to trust the application of this system to nanomaterials.

The broad social and cultural accounts provided by the previous scholars are useful for evaluating the social construction of risks in society. However, for narrow cases such as nanotechnology it is also important to look with more specificity at how risk arguments are applied in public policy. One philosophical perspective for understanding how risk arguments may be developed in the policy arena is presented by Kristin Shrader-Frechette. She argues that social reductionists such as Douglas and Wildavsky overemphasize the role that societal/cultural values actually play in risk assessment, while scientific reductionists or "naïve positivists" (whose view could be interpreted as being reflected in the EPA definition of risk cited earlier) underemphasize both ethics and democratic procedure in risk assessment.[22] Shrader-Frechette labels her own approach *scientific proceduralism.* It is procedural in that the process of risk evaluation is guided by ethics and democracy; it is scientific because it is logical, uses probabilities, and purports to explain and predict hazards.[23] To operationalize this approach, the author recommends that several risk analyses be done for a given problem, using different methodologies so that we see "all sides of a given 'story'."[24] Like the previous authors, Shrader-Frechette recognizes that society's view of risk ought to be enriched beyond technocratic risk analysis; unlike Beck and Douglas and Wildavsky, she appears relatively unconcerned about keeping risk within the dominion of science. However, I would argue that getting all sides of a story is not sufficient if the given story is only a slice of a broader narrative that is not given adequate treatment in policy debates. Shrader-Frechette's scientific proceduralism may indeed be useful in enhancing the robustness of risk analyses. Nevertheless, just doing risk analysis is not a sufficient exploration of environmental and societal impact.

Sheila Jasanoff and Anne Chapman explore difficulties that arise through our current approaches to risk analysis. Jasanoff labels the use of risk in the policy arena as "trans-scientific" in that it is coextensive with a regulatory science framework that involves producing and synthesizing knowledge for the purpose of making risk predictions that inform policy decisions.[25] Because the kinds of risk predictions called for in the policy arena are linked to the legal defensibility of regulatory decisions, how science is used to produce and characterize health and safety findings is bound up in the legal frameworks and institutions that are central to liberal democratic governance. For Jasanoff, it is the inherent uncertainty of scientific information that binds science and policy together in a trans-scientific articulation of risk.[26] This is my observation as well. Because of the complexity of biological systems and confounding factors such as genetic variability and differential susceptibility to disease, hypotheses on whether any given chemical substance causes one health effect or another are almost always severely underdetermined by the evidence acquired from laboratory animal studies or epidemiology. This is one important reason why, to my knowledge, science has never resolved a risk debate on any chemical substance and therefore is not likely to settle nanotechnology risk issues.

Anne Chapman argues that a way to avoid intractable risk debates is to shift the focus to the *riskiness* of the technology under evaluation. Whereas estimates of risk are tied to outcomes associated with a technology's interaction with people or the environment, riskiness relates to the inherent properties of the technology itself, or what Chapman calls the "epistemic possibility of harm."[27] For instance, a typical risk evaluation of a chemical might involve establishing a causal relationship between human exposure to the chemical—say, ingesting the chemical by eating an apple containing pesticide residues—and that exposure triggering an adverse health effect. The risk, then, occurs because of situations arising "around" the chemical—

in this case, by eating the apple. Strategies for addressing the risk therefore tend to focus on the situation, such as washing the fruit or buying organic produce. By contrast, Chapman's riskiness approach entails focusing on the chemical itself, irrespective of what situations may arise that lead to the chemical's encounter with biological systems. Is the chemical likely to bioaccumulate in tissues? Will it persist and be mobile in the environment? Or will it break down into benign constituents? Strategies for addressing riskiness therefore focus on the technology itself, and shift responsibility for safety to those who control the technology rather than to those who are impacted by it. Also important to the concept of riskiness is its shift in the burden of proof in determining the acceptability of a technology. Typically, risk assessments focus on what knowledge is available on hazard and exposure, sometimes with an implicit assumption that no data equates to no risk.[28] However, for a riskiness determination, Chapman argues, lack of knowledge increases a technology's inherent riskiness because we cannot judge the inherent harmfulness of the technology.[29] Jasanoff, Shrader-Frechette, and Chapman all argue for balanced analyses that acknowledge the limits of science and the important role that politics plays in societal discourse on risk issues.[30]

The literature on risk suggests that both social construction and cultural/institutional factors place boundaries around the possibilities for risk discourse in the United States. Beck provides a vision for enlarging boundaries, but the more closely risk scholarship focuses in on the application of risk evaluation as an aspect of governance in liberal democratic capitalist society, the less optimistic one feels about the potential for such boundary movement. If this were the end of the story, the possibilities for enlarging that discourse would seem rather bleak. But we are nowhere near the end of the story. In the next section, risk is put to work through regulatory science. It should be clear through this discussion that regulatory science, as currently practiced, will not serve nanotechnology any better than it has addressed industrial chemicals in terms of enriching risk discourse. However, it should also be apparent that the inadequacies of regulatory science are made obvious through the nanotechnology case, and bringing these inadequacies out into the open presents opportunities for change.

REGULATORY SCIENCE

When people employ science to make risk arguments in a policy context, they are entering the realm Sheila Jasanoff calls regulatory science.[31] In this arena, scientific data are marshaled to support positions on whether and how government should regulate social activities. Jasanoff partitions regulatory science into three types of activities. *Knowledge production* is the generation of scientific information. This information may come from various sources, including academic institutions, industry, and government agencies themselves. The collection and analysis of this information in order to develop some characterization of its meaning for a particular decision context is *knowledge synthesis.* Jasanoff's statement that the products of this activity are "rarely innovative and may never be submitted to the discipline of peer review and publication" may indeed apply to some routine uses of science, such as the licensing of derivative products for which there already exists a large body of data.[32] However, we shall see that this is not necessarily the case for nanotechnology regulatory science. Jasanoff's third category of regulatory science is *prediction,* in which the synthesized information is used to estimate the probability of an event or outcome. [33] For the regulation of chemicals, the outputs of prediction activities typically are quantitative assessments of the risk of some effect from exposure to the chemical. For instance, a risk assessment for air pollutant X may estimate a probability of air pollutant X causing a certain number of new cancer cases per year for people who live within a certain distance of a facility emitting air pollutant X.[34] Because risk assessment is central both to environmental regulatory science generally and nanotechnology specifically, it warrants additional discussion.

Risk Assessment

For environmental regulatory science, risk assessment involves developing probabilities of harm based on estimates of how much of a contaminant is present in the environment, how toxic the contaminant is, and how much a person, animal, or ecosystem is exposed to the contaminant. Government agencies have been assessing the risk of chemicals and other substances for decades. However, it was not until after the National Academy of Sciences published its 1983 report, *Risk Assessment in the Federal Government: Managing the Process* (often called the "Red Book"), that the United States, led by EPA, began to develop formal guidelines for environmental risk assessment.[35] In 1984 EPA published *Risk Assessment and Management: Framework for Decision Making,* and since that time has published numerous guidelines and technical guidance manuals on specific aspects of risk assessment.[36] Today, few decisions on the environmental impacts of chemicals or other contaminants are considered complete without the inclusion of some form of risk assessment.

The database to support a risk assessment consists, broadly speaking, of hazard and exposure data. Although human clinical or epidemiological data on rare occasions are available, toxicology studies on laboratory animals or organisms in cell cultures typically are used to assess the potential hazard from environmental contaminants. Animals or cells are dosed at various levels until adverse health effects are observed, and mathematical models are used to extrapolate the results from the test organisms to humans (or, in ecological testing, to animal or plant populations that are deemed to be indicators of ecosystem health). To estimate potential human or ecological exposure to a contaminant, either actual measurements are taken or computer models estimate contaminant levels in air, water, soils, or food. These hazard, effects, and exposure data are analyzed and a characterization of risk is made. Uncertainty is an inherent property of all risk assessments.[37] Rarely are databases complete and where there are gaps, risk assessors may apply default values or uncertainty factors. Even when a database is quite complete (*entirely* and *permanently* complete probably being impossible, given the continual advance of scientific understanding), results and findings are subject to different interpretations.

Using Regulatory Science

Government agencies apply information generated through environmental regulatory science in ways that correspond to the specific authorities granted to those agencies for the administration of laws governing public health, safety, and environmental protection. Statutes covering environmental, health, and safety protections vary within US governmental jurisdictions and among nations, but within western democracies the way in which regulatory science is used is basically the same. Regulators use a risk assessment (from Jasanoff's "prediction" stage of regulatory science) as a chief supporting document to justify a decision. Depending on whether or the extent to which a particular statute permits the consideration of benefits, the risk assessment may be incorporated into a larger benefit-risk assessment. Generally, the assessment is characterized with respect to the quality, completeness, and relevance of the scientific information, with discussion of the uncertainties in the assessment's estimates of risks and benefits.[38]

There are various types of actions to which environmental regulatory science may be applied. Product licensing decisions, such as the granting of an EPA registration for a pesticide, use regulatory science information generated primarily by the company requesting the registration, although this information often is supplemented by information either developed within the government or found in the published literature. Standards for air and water quality also rely on regulatory science information. While some of this information may come from private industry, the bulk of it is produced by government laboratories or academic institutions (many of which are funded through federal grants for the study of pollutants and their effects). Regulatory science also is used to evaluate new chemicals coming into commerce, or existing chemicals that have triggered some risk concern or have been included in legally mandated review processes, such as the requirement that EPA produce a Candidate Contaminant List of possible water pollutants.[39] In addition to statutorily mandated activities, regulatory science is used to inform non-regulatory activities, including research agendas. For instance, in addition to being a generator of regulatory science information that is used by agencies such as EPA, the United States Geological Survey (USGS) uses regulatory science information to guide some of its programs. An example is its Toxic Substances Hydrology Program, which identifies its priorities for studying water contaminants based in part on information from federal regulatory agencies such as EPA.[40] The information generated by USGS in turn is used to inform EPA regulatory decisions on water contaminants, thus completing a circle whereby regulatory science informs policy needs which then shape the research agendas of non-regulatory entities, which generate scientific information that feeds back into regulatory decision making.

Guidance on risk assessment and risk characterization, such as that developed by EPA, acknowledges that information from regulatory science is but one input into environmental policy decisions. In reality, however, decision documents rarely give other factors—such as politics, economic considerations (when a benefits assessment is not included), and social equity—much if any prominence. Even if such considerations are mentioned in a decision document, one would be hard pressed to find an instance where those considerations are more than a footnote to the centerpiece risk or benefit-risk assessment.[41] I believe this is the case because regulators want to make sure their risk cases stand up in court, and there is a strong need within American administrative and legal cultures to establish a quantifiable relationship between risk reduction and the costs and/or benefits of taking a regulatory action. Even when EPA indicates that is considering welfare effects, as it did in its 2009 greenhouse gas endangerment finding, what it means by "welfare" are those aspects of risk, benefit, or cost that are amenable to quantification, even if the particular finding does not do so.[42]

Regulatory science has a symbiotic relationship between scientists and policy makers: policy makers need scientists to generate information to support environmental decisions; scientists need policy makers' problems as sources of work and, in many cases, funding. Science and policy making may be a forced fit, as Jasanoff and Chapman argue, but it is a coupling that serves to sustain—and to an extent, legitimize—both activities. This relationship supports the argument that the use of science, in particular regulatory science, occurs within a broader context of ordered, calculated governance. Regulatory science performs three functions—simplification, co-production, and discipline—that give it the character of what Peter Miller and Nikolas Rose label a *technology of governmentality* operating within a calculating democracy.[43]

Simplification

Regulatory science serves to simplify complex chemical-biological interactions into numerical expressions of risk. James Scott, in his case studies on European land management, points out that governments must apply these simplifying approaches because the complex and nuanced interactions at the local level (or, for the environment, in nature) are not translatable to the bureaucratic needs of the state.[44] Scott gives an example of how defining a bushel of wheat was troublesome for a new French national government attempting to establish commodity trading standards, because there existed throughout the country a heterogeneous mix of village basket weavers who made bushel baskets in different ways and of varying sizes. The solution was for the government to establish a standard size and measurement approach for a wheat bushel. For present-day chemical (and nanoparticle) risk assessment, an example of such simplification is establishing exposure factors for water consumption. An average American is constructed for exposure assessment purposes. An adult is assumed to consume two liters of water a day; children weighing less than ten kilograms are assumed to consume one liter a day. Standards for the "average American" are also set for body weight (70 kg) and a number of other factors.[45]

Another type of simplification through calculation relates to Porter's observation that we quantify an inherently nonstatistical Nature in order to manage the world.[46] In assessing the risks of chemicals we do such simplification in estimating effects to humans or individuals of other species. We conduct laboratory animal or cell culture studies to estimate the probability that exposure to a particular chemical will cause some additional number of cancer cases to a population of people potentially exposed to the chemical. This simplification is expressed in such terms as Chemical X may cause one additional cancer case per one million people exposed over their lifetimes. Do we really believe that such probabilities are a true reflection of what occurs in nature as chemicals interact with biological systems? No, we are not that naïve. Yet we proceed with the simplification because it is useful for the governance purposes to which regulatory science is employed.

Co-production

Regulatory science owes its existence to the co-production of scientific activity and environmental decision making. Jasanoff describes how co-production—the idea that how we view the world is contingent on how we live in the world—occurs along four possible pathways: making identities, making institutions, making discourses, and making representations.[47] In our calculated democracy, science (and other calculating endeavors) and governance are co-produced along each of these pathways. Ezrahi shows how the identity of science is transformed as an instrument of democratic politics, such that science becomes part of the public performances in which democratic governments must engage to demonstrate their transparency and accountability.[48] Science and engineering build up an entire cost/risk-benefit infrastructure, which in turn is reflected in congressional analyses, debate, and decision making, all made possible through regulatory science. Rose discusses the discursive pathway when he describes numbers fostering a rhetoric of disinterest.[49] The political need for objectivity and alienation of the political actor from his or her actions leads to the "technicization of politics" as well as to the reinforcement in science of the Mertonian norm of disinterest.[50] For instance, risks characterized in terms of calculated probabilities expressed numerically are not accompanied by fuller descriptions of cancer, such as the anxiety one experiences when presented with a cancer diagnosis, the pain experienced through disease progression and treatment, or the disruptions that occur within the cancer patient's family or other support networks. Indeed, even the prominence of cancer as a health effect of concern is an example of co-production, in that, as Bosso illustrates, the strong emphasis on estimating the carcinogenic potential of chemicals arose out of the "cancer scare" episodes and subsequent congressional hearings of the 1950s.[51] Finally, Ezrahi discusses *representations*: realities created from the co-production of science and politics, and projected into society.[52] In conducting assessments and calculating probabilities, regulatory science creates a particular numerical representation of the consequences of interactions between contaminants and biological systems.

Discipline

Regulatory science is one means through which our calculating democracy imposes rhetorical and behavioral discipline on those who govern as well as on those who are governed. Porter observes that to accurately measure, you have to discipline people as well as instruments.[53] *Rhetorical discipline* means that measurement instruments must provide values that fit within the boundaries of societal discourse, and policy makers and citizens alike must be able to think and speak in the languages of those values. For example, across American society there is an understanding of the term *household income*. A richer discussion of *wealth* would be unmanageable and subject to multiple interpretations, whereas the household income statistic provides for disciplined national discourse. Likewise for regulatory science, language is disciplined to fit within the context of risk of disease or direct ecological impact; for instance, cancer in humans, or the reduction of species diversity. As a result, these are the impacts for which we design assessment tools to measure and construct language to describe.

Behavioral discipline relates to ensuring order in politically transparent and accountable societies. Ezrahi identifies three principal roles for science in supporting his *instrumental* concept of discipline: (1) reconciling freedom and order, (2) depersonalizing political power while retaining responsibility of the agent, and (3) making agents accountable to the public for actions taken on the public's behalf.[54] Vaccination programs, which employ a type of regulatory science, are an example of how medicine serves as a disciplining instrument of public health policy. First, a sound public health strategy necessitates imposing some order on society while acknowledging personal freedoms. (If you choose not to vaccinate your children, they will not be allowed to attend public school.) Second, findings from medical science direct policy makers to institute specific vaccination programs. Third, follow-on epidemiology and other tracking activities provide indications of vaccine efficacy and side effects, and this information is placed in the public domain to allow for government accountability on the safety and effectiveness of the program. In this manner, regulatory science provides a legitimizing function to government mandates by making present for public evaluation the effectiveness and value of such mandates. This notion of accountability helps explain the importance that government agencies place on publishing their major risk assessments for public comment in the *Federal Register*.

Trust and Risk

Within the context of expert systems such as regulatory science, Giddens sees trust as confidence in the reliability of the expert system, with that confidence contingent on faith in the correctness of the knowledge that underpins the expert system.[55] Trust helps establish boundaries around what we consider to be an acceptable risk. If we trust that regulatory science, as an expert system, will generate risk-related information we can believe in, then for nanotechnology we will have confidence that we can calculate the degree or probability of risk that a nanoparticle presents.[56] It also increases our confidence in being able to manage the risk.

However, regulatory science is a weak expert system, and risk analysis is a mistrusted discipline. The weakness of regulatory science stems from the strong underdetermination by the evidence that can be assembled to formulate hypotheses of causality. As Jasanoff points out, science is ill equipped to provide conclusive evidence in support of decisions that by their nature are adversarial and therefore under liberal democracy are subject to adjudication.[57] Science does not offer proof: it constructs statements in support of hypotheses which, as history has repeatedly demonstrated, will at some point be modified or refuted outright. Yet regulatory systems, particularly those in the United States, operate within an adversarial model modeled on, and underpinned by, the judicial system, and science is expected to help provide evidence in courtroom arguments about regulation.

Regulatory science, fraught with epistemological underdetermination, is poorly suited for bringing resolution to legal disputes.

Environmental risk analysis is mistrusted for at least two reasons. First, the disciplines that produce the regulatory science information have an extremely difficult time describing with any degree of precision how contaminants behave in the environment. At every point—from estimating releases to modeling pollutant transport to estimating impacts on a biological receptor—uncertainty builds upon uncertainty as models and measurements attempt to capture static information in dynamic systems. Second, risk analysis information is used inconsistently by proponents of various regulatory positions. As the significant body of hearing testimony on pesticides risk shows, those who support regulation of a substance based on high risks identified in an assessment will point to the "proof" in the risk assessment that the substance is dangerous; those who oppose regulation will point out the uncertainties in the assessment. However, should an assessment of another substance give an estimate of a low probability of risk, those same advocates of regulation in the first case are quite likely to point out the deficiencies in this case's assessment, while those groups who opposed regulation in the first case will likely in this case tout the strength of this assessment's "safety finding." These scenarios play out repeatedly in nearly every environmental risk case for which regulatory decisions are contemplated.[58] It is unclear whether anyone, other than those who generate risk assessments or provide the information that supports them, really trusts the conclusions reached in risk assessments, or whether they have become merely the best tools available to support positions on regulatory decisions. Even if they are the best available tools, that does not make them adequate for fully explicating environmental risks.

I would argue that in part because of its weakness as an expert system, risk assessment practitioners and advocates seek to bolster risk assessment's acceptance as a scientific discipline by relying heavily on quantification. Porter observes that in cultures where experts occupy a strong social position and their opinions are trusted, such as in French society, quantification is less relied upon as a justification for decisions than it is in countries such as the United States, where experts occupy a relatively weak social position and society places greater trust in legal processes than in expert knowledge or the positional authority of individuals.[59] Porter also argues that disciplines with weak predictive capability, such as those in the social and biological sciences, latch onto the use of statistics as a means to convey the perception that they are strong in their inferential capabilities.[60] For regulatory science, projecting authority through quantification is essential for it to play a meaningful role in our calculated governance.

Regulatory science is a technology of governmentality insofar as it serves as government's means for identifying and articulating a particular notion of risk constructed within our calculating, disciplined liberal democratic society. However, it is a weak technology, or expert system, and as such it is not only inadequate for fully discussing and addressing environmental risks, but because of its weaknesses will not, if left alone, maintain the trust of American society as it confronts—and is expected to address—increasingly complex interactions between new technologies and the environment.

REGULATORY SCIENCE FOR NANOTECHNOLOGY

Companies are developing and producing nanoscale objects, particles, and materials because nanotechnology adds performance characteristics that make products better and therefore provides competitive advantages in the marketplace. However, some of the properties that make nanoparticles useful are the same properties that raise concerns about nanoparticles' possible risks. For example, largely because of their small size, nanoscale titanium dioxide particles scatter very little visible light and therefore are transparent when applied on skin. This characteristic has made them popular, replacing some types of sunscreens that, while protective, appear white on the skin. However, this same beneficial property—small size with accompanying high reactivity—has raised questions about whether the nanoscale particles can be absorbed through the skin and possibly cause cell damage that could trigger the onset of disease. While different nanomaterials have material- or particle-specific beneficial properties with concomitant property-related risk concerns, the basic nexus between unique properties, new benefits, and risk concerns is central to nanomaterial environmental, health, and safety issues. Regulatory science seeks to understand whether any relationships exist between the properties of specific types of nanoparticles and adverse environmental impacts and, if such relationships exist, what can be done to minimize the potential for harm. For instance, some studies have suggested a relationship between both particle size and electrical charge and a particle's ability to cross the blood-brain or placental barrier.[61] The understanding of such a relationship could lead to redesigning the particle to change either its size or charge, while retaining its beneficial performance characteristics, thus maintaining the particle's benefit but eliminating at least one aspect of potential risk. In applying regulatory science to understand interactions between nanomaterials and the environment, and how particle/material properties affect those interactions, EPA's *Nanomaterial Research Strategy* asks four major questions:

- What nanomaterials, in what forms, are most likely to result in environmental exposure?

- What particular nanomaterial properties may raise hazard or exposure concerns?
- Are nanomaterials with properties of concern likely to be present in the environment in concentrations of concern?
- If the answer to the above question is "yes," what can be done to reduce the material's potential to create risk?[62]

While taken from an EPA document, these questions are in essence the same ones asked by regulatory science organizations around the world.[63] What follows is a by-discipline overview of the regulatory science activities in which scientists are engaged to address these four questions.

Material Characterization

An important prerequisite to producing nanomaterial test data that are understandable by the rest of the scientific community is describing in detail the material that one is testing. Essentially, characterizing a nanoparticle means understanding and communicating what it looks like, what it's made of, and how it interacts with its surroundings.[64] Doing so has proven to be extremely challenging. First, there are many forms of nanoparticles, even within the same class of materials, that are fabricated using a variety of processes. For example, there are various forms of multi-walled carbon nanotubes, produced through different fabrication processes. Some of the tubes are straight, and others are bent and tangled; some are nearly pure carbon, while others contain high percentages of metals and other contaminants that are by-products of the production process; and some are less than 100 nanometers long, while others may be many microns (one micron equals 1,000 nanometers) in length. Even within one specific type of tube, they are never all the same size, but rather any given sample will contain a distribution of different-sized particles within a certain range. How a particle is made will also determine such characteristics as its electrical charge, solubility, and stability. Without clear understanding and communication of a particle's characteristics, it would be difficult to determine what properties may be responsible for one material being more or less toxic than another.

Research activities in this area include identifying the critical characterization parameters for the various tests that are required to provide data for risk assessments, agreement on how to measure those parameters—such as what equipment to use, how to calibrate the instruments, and how to prepare materials for measurement—and how to account for variations in materials and impurities or other materials that may be incorporated into the nanoparticle. A number of different types of sophisticated equipment are used to characterize nanoparticles. Transmission electron microscopy, scanning electron microscopy, confocal microscopy, dynamic light scattering, and atomic force microscopy are some of the typical approaches used to measure and create images of the particles.[65] These approaches use expensive, typically large pieces of equipment that are housed in laboratories, often under special environmental conditions to ensure measurement consistency. Scientists must have specialized training in operating each of the different instruments.

Environmental Fate, Transport, and Transformation

An important question is *when* to characterize nanoparticles. For those who look at the fate, transport, and transformation of nanoparticles, the answer is to understand the particles as they are released into and move through air, water, soil, as well as through biological systems. Because the particles typically are highly reactive and tend to agglomerate (clump together) or attach to other materials, they likely will change significantly once they are released into the environment. The development of techniques to test nanoparticles in environmental media, both in the laboratory as well as *in situ,* has proven to be a major challenge for researchers.[66]

Finding nanoparticles in environmental media, let alone tracking them, is at best a difficult undertaking. It is worse than looking for a needle in a haystack—it's like trying to find a needle in a stack of hay and needles mixed, since particles at the nano scale also occur naturally in the environment. Scientists hope that techniques such as tagging particles with radioactive isotopes or using fluorescing markers will help find and track particles as they move through the environment.[67]

Recognizing that using direct observation to understand the environmental behavior of the many various types of nanoparticles may not always be tenable, models also will be needed to predict nanoparticle behavior. Fate and transport models are used today for many contaminants. For example, models predict the movement of mercury all the way from its emission from coal-fired electrical power plants; as it is carried through the air; and as it precipitates out of the air and into water bodies, transforms from elemental mercury into methyl mercury, and then moves up the food chain as it bioaccumulates in animals' fatty tissues. Research is now investigating how these same models might be modified to predict the fate, transport, and transformation of nanoparticles in the environment. The models' outputs then must be compared against observational data, which to be generated will require the material characterization approaches discussed in the previous section.

Ecological and Human Health Effects

Understanding whether nanoparticles have the potential to harm people or other living organisms is perhaps the most-prominent aspect of nanotechnology regulatory science. In ecological effects research, scientists attempt to understand whether populations of animals or plants may be adversely impacted by contaminants or other stressors. Relevant biological systems used for testing ecological effects reflect decisions made regarding what species serve as indicators of ecosystem health, such as fish species, aquatic grasses, or insects.[68]

For humans, the concern is whether an individual—albeit a random individual within a population of potentially exposed people—may suffer ill effects from nanoparticles. Researchers attempt to understand how nanoparticles might be absorbed into the body, where in the body they may go if absorbed, and how they interact with the body and whether such interactions may trigger biological events that cause adverse health effects.[69]

Scientists go through a hazard identification process to understand what health effects, including but not limited to cancer or birth defects, might result from exposure to the nanoparticle in whatever system the particle is being tested. A dose-response assessment is then conducted to describe how the adverse health effects identified in the hazard assessment relate to how much, and under what conditions, the test systems are exposed to nanoparticles. Estimates are then made about the likelihood of those effects being seen in humans under probable exposure scenarios.[70]

While in the future clinical or epidemiological data on humans may be available, at present very little if any such data exist for nanomaterials. Therefore, the bulk of nanoparticle health effects information is generated *in vivo* with laboratory animals or *in vitro* using cell cultures. Various models that were developed for industrial chemicals then are used to extrapolate laboratory findings on nanoparticles to potential human effects, with uncertainty factors applied to the estimates to account for interspecies extrapolation, susceptible human subpopulations, and test data gaps.[71]

Exposure

As nanomaterials are incorporated into increasing numbers of products, at some point they are going to enter the environment, possibly resulting in exposures to humans and other living things. Potentially exposed humans include workers exposed during the production of nanoparticles and their incorporation into products, as well as in the disposal or recycling of the materials. The general population may be exposed either through emissions during production, disposal and recycling, or through use of products that contain nanomaterials.[72] Other species might be exposed through the same activities, as the materials are made, used, and reused or disposed of, and as particles are emitted into air, water, and soil.

As with other nanotechnology regulatory science disciplines, developing detection and measurement technologies is a high priority for exposure research, as is the development of models of how particles might reach humans or other receptors.[73] With heavy emphasis given to toxicology and effects research, exposure science has until recently received relatively little attention. However, recognition that it likely will take years to develop robust databases on nanoparticle health effects has stimulated interest in trying to understand whether, and if so under what scenarios, exposure may occur. Through modification of existing technologies to detect particles incidentally produced through industrial processes, progress has been particularly significant in detecting airborne nanoparticles in the workplace.[74]

Exposure to nanoparticles at sites within the human body or in other biological systems is also a focus of research. Fundamental to this research is determining the biologically effective dose, which is the amount of nanoparticles that reaches organs, cells, or other sites in concentrations that are sufficient to cause an adverse effect.[75] Crucial to conducting such research is the ability to detect, describe, and measure particles as they move through biological systems, and therefore this research depends upon research on material characterization and fate, transport, and transformation.

What the above research activities demonstrate is the reliance of nanoparticle regulatory science on instruments and models for observing, predicting, and interpreting the significance of nanoparticles in the environment or in living systems. Just as how, in Beck's risk society, the risks of science are given back to science to address, so too in the nanotechnology case does society hand over to technology—in this instance the operators of complex imaging devices and computer models, and those who interpret their outputs—the task of understanding technology's unintended consequences. Joseph Pitt argues that the outputs of imaging devices really are not images of nanoparticles, but rather are heuristics or extended metaphors for what we imagine nanoparticles to be.[76] Pitt contends that what these tools produce is not a truly realistic image of a nanoparticle, but rather is a computer-generated composite description of the features of the particle that the device encounters, such as by an electron beam bouncing off the surface of a particle and recording the impulse and manipulating it with a computer program, that is color enhanced and presented as an image of what the nanoparticle looks like. One of Pitt's concerns is that by representing these pictures as images, the scientific community presents a simplified picture of how phenomena occur at the nano scale as well as gives the impression that we know more than we actually do about how nanoparticles behave.[77] In fact, as best we understand, quantum effects are highly dynamic and individual events are unpredictable. This environment of constant motion and random behavior cannot be captured in a TEM picture. Pitt's point, then, is that pictures of nanoparticles lend undeserved authority to epistemological claims by scientists regarding what they know about activity at the nano scale, and present the public with the misconception that we understand the behavior of particles at that scale. If Pitt is correct, then as nanotechnology regulatory science proceeds with generating and compiling data, it is mounting uncertainty upon uncertainty as information is integrated within and across disciplines to formulate characterizations of nanotechnology risks. Even if Pitt were not entirely correct, and if validation across imaging techniques does lead to some sort of consensus on what a nanoparticle "looks like," we are still left with significant underdetermination due to what information or activities the images and data *do not* convey, as well as to the inability of static pictures or readings to predict how the behavior of dynamic systems changes as they move through time and space. To some degree these concerns also arise with attempts to measure other contaminants in the environment. What is different about nanoparticle images is that, compared to chemical characterization outputs, their detailed and often colorful pictures tend to convey a sense of precision and understanding that cannot be reconciled with our lack of comprehension of particle kinetics and quantum-level behavior.

Risk Assessment and Risk Management of Nanomaterials: A New Direction

While risk assessment remains the standard approach for characterizing the potential for harm of traditional chemicals, some scientists involved in generating and using regulatory science for nanotechnology are moving down a different path. That path closely follows Chapman's riskiness approach, in that these scientists are focusing on understanding what properties of nanoparticles may cause harm to organisms, and how those properties may be modified or otherwise addressed so that a particle may be made safer while retaining the beneficial performance characteristics that the particle was fabricated to achieve. For example, a few studies suggest that certain types of multi-walled carbon nanotubes—those that are long, thin, straight and durable, similar to asbestos fibers—when artificially instilled into the abdominal lining of laboratory animals induce effects similar to those observed in the early stages of asbestos-related mesothelioma.[78] A question for decision makers is how to use this information on particle property-induced effects to encourage the fabrication of safe multi-walled carbon nanotubes. Strategies developed by governments, industry, and environmental NGOs advocate this approach.[79] With this focus on material properties, the concepts of assessing and managing risk (or riskiness) become coextensive and, as I argue below, have consequences for how the imbrication of risk and trust articulated by Giddens plays out in both the construction and use of regulatory science for nanotechnology.

Documents such as the Environmental Defense (ED)-DuPont *Nano Risk Framework* and EPA's *Nanomaterial Research Strategy* suggest uneasiness with relying solely or even primarily on risk assessment to inform decisions on the environmental safety of products. There is a general recognition that performing risk assessments for nanomaterials is not only going to be a time- and resource-intensive process, but also that such assessments are not going to lead to consensus on whether or not any given nanomaterial is safe. Both the ED-DuPont and EPA documents recommend looking at the inherent properties of the nanomaterials themselves—to identify features that may, as described by Chapman, make them risky—as a starting point for deciding if, and if so, how, the material should be introduced into products.[80] These approaches are a conceptual advance over how regulatory science has addressed traditional chemicals, but whether they succeed in practice may depend on whether this new thinking also leads to trust among actors that such modification to the regulatory science expert system will yield acceptable decisions on nanomaterial safety.

Trust will need to be accompanied by an increased tolerance for uncertainty among stakeholders in the decision process, in particular by manufacturers and companies that incorporate nanomaterials into their properties. Consider the example cited above of the long, straight, and durable multi-walled carbon nanotubes. Under highly artificial test conditions, these materials have produced effects similar to those seen at the onset of disease associated with asbestos exposure. The studies suggest that the effects are directly related to the nanotubes' properties; specifically, that they are long, straight, and durable.[81] Shorter nanotubes do not elicit the same adverse effects. Logic would suggest that—without waiting years for a complete toxicological database and the modeling of multiple exposure pathways across product lifecycle stages, which would be needed under a full quantitative risk assessment—a conceptually straightforward approach to reducing the riskiness of this type of multi-walled carbon nanotube would be to find a way to make shorter or less-durable multi-walled carbon nanotubes that retain the beneficial performance characteristics of the long ones. However, doing so would entail a business cost to carbon nanotube fabricators. The business community's willingness to incur such an expense likely would depend on their trust that the regulatory science information is valid and that it will be incorporated into regulators' decision analyses. The regulators, in turn, must trust both the science as well as industry's commitment to seek solutions for safer materials. Society at large, including public-interest advocates such as environmental NGOs, will have to believe that this emphasis on identifying and addressing problematic particle properties will lead to safer nanotechnology products. This is much to ask of a society that heretofore has taken a legalistic, adversarial approach to settling—or more often, not settling—disputes over chemical risks.

Such an approach also runs counter to the demands of calculated governance, in particular by elevating the prominence of individual human agency in the decision process. Risk assessments foster in two ways what Ezrahi labels democratic instrumentalism: first, they depersonalize risks by making them quantitative scientific constructions used by, but independent of, individual public officials; second, they provide a transparent and adjudicable product (the risk assessment document) that the public can use to hold government accountable for its decisions.[82] The nanomaterials risk management approach described above upsets democratic instrumentalism by forcing individuals to take actions without the support of quantified assessments of risk. This runs counter to the dominant public policy discourse for chemicals, which does not lend itself to non-quantified approaches to environmental risks. It remains uncertain whether the current direction of nanotechnology regulatory science offers the possibility for such change.

CONCLUSIONS ON NANOTECHNOLOGY RISK

Regulatory science, as manifest in chemical risk assessment, falls short on its own as a policy-making tool because it is both undemocratic and technically inadequate. It is undemocratic not because it is an expert system—the public needs its governance institutions to employ expert systems. Rather, risk assessment is a system so inherently weak and fraught with epistemological underdetermination that the experts who create risk assessments hold them together with quantitative uncertainty factors, extrapolations, and modeling assumptions—the methodological equivalents of wire and duct tape—to such an extent that the construction and characterization of the assessment's risk estimate are opaque not only to the public but to the policy makers who must decide how such assessments inform their decisions. This lack of transparency is increased in the nanotechnology case because of the complex devices needed to detect, measure, and image nanoscale particles, as well as because the detailed pictures of nanoparticles generated by electron microscopy give the misleading impression that the particles are static and highly controllable, and do not reflect our very poor understanding of nanoparticles' behavior in the environment.

There is little about regulatory science in its current state, given its great uncertainties and lack of transparency, to foster public trust in governance institutions' ability to identify, avoid, or manage potential environmental risks from emerging technologies. At the same time, for nanotechnology we see government boosterism in the form of the NNI at the same time that government struggles to apply regulatory science approaches to nanomaterials, reflecting society's appetite for new, but safe, products and sources of employment and wealth. Chapman's riskiness approach suggests potential ways to address these potentially contradictory societal needs through risk-avoidance approaches such as green chemistry, in which environmental sustainability is engineered into product design.[83] However, one should view with caution this perspective, consistent with Hajer's ecological modernization, that environmental sustainability can be achieved by technocratic means.[84] Harkening back to Beck's reflexive modernization, society has a poor track record of enhancing environmental protection by turning the solution of science- and technology-created problems back over to science and technology to fix.

That said, the nanotechnology case demonstrates that the role of risk assessment and regulatory science in environmental debates is not an all-or-nothing proposition. Green chemistry and other design-focused approaches push discussions of risk more upstream in material- and product-development decision making, and such earlier consideration of environmental impacts would be a welcome change. Also, the very difficulties in applying risk assessment approaches to nanomaterials in such fundamental areas as material characterization may lead to increased awareness among parties involved in environmental governance that while risk assessment remains useful for informing decision making, its utility was perhaps never as high as once thought and apparently is even less so with complex new materials.

Questions remain as to whether, despite its inadequacies, there is room for more than regulatory science in environmental risk discourse. If so, how much room is there? Can regulatory science be de-centered in risk debates and if so, how? Prior to taking on the *how* question, going beyond the technocracy of risk assessment and regulatory science into further exploration of the nanotechnology debates will reveal deeper societal forces acting on discourse. In the next chapter, exploring language provides a useful means for beginning to identify and understand those forces.

NOTES

1. http://www.epa.gov/ncea/iris/subst/0144.htm.
2. US Government Accountability Office, *Chemical Regulation: Observations on Improving the Toxic Substances Control Act,* GAO-10-292T, Washington, DC: 22 Jan 2009, 2.
3. US Environmental Protection Agency, *Risk Assessment Principles and Practices,* EPA Document EPA/100/B-04/001, March 2004, 2.
4. Ulrich Beck, *Risk Society: Towards a New Modernity*. London: Sage Publications, 1992, 21.
5. Ulrich Beck, *Risk Society: Towards a New Modernity*. London: Sage Publications, 1992, 13. This use of *reflexive* seems consistent with David Bloor's articulation of the word, which in the Strong Program context means that just as sociology has a legitimate role in examining science, so too is it legitimate to use sociology's "patterns of explanation to examine sociology itself." See David Bloor, *Knowledge and Social Imagery*, Second Edition (Chicago: The University of Chicago Press, 1991), 7.

6. Beck, *Risk Society*, 23.
7. Beck, *Risk Society*, 167.
8. Beck, *Risk Society*, 23.
9. Ulrich Beck, *Ecological Enlightenment: Essays on the Politics of the Risk Society* (Atlantic Highlands, NY: Humanities Press), 9.
10. Ulrich Beck, *Ecological Enlightenment: Essays on the Politics of the Risk Society* (Atlantic Highlands, NJ: Humanities Press), 24.
11. Ulrich Beck, *Ecological Enlightenment: Essays on the Politics of the Risk Society* (Atlantic Highlands, NJ: Humanities Press), 29.
12. Ulrich Beck, *Ecological Enlightenment: Essays on the Politics of the Risk Society* (Atlantic Highlands, NJ: Humanities Press), 46.
13. Ulrich Beck, *Ecological Enlightenment: Essays on the Politics of the Risk Society* (Atlantic Highlands, NJ: Humanities Press), 158.
14. Mary Douglas and Aaron Wildavsky, *Risk and Culture: An Essay on the Selection of Technical and Environmental Dangers* (Berkeley: University of California Press, 1982), 7.
15. Mary Douglas and Aaron Wildavsky, *Risk and Culture: An Essay on the Selection of Technical and Environmental Dangers* (Berkeley: University of California Press, 1982), 14.
16. Mary Douglas and Aaron Wildavsky, *Risk and Culture: An Essay on the Selection of Technical and Environmental Dangers* (Berkeley: University of California Press, 1982), 73.
17. Mary Douglas and Aaron Wildavsky, *Risk and Culture: An Essay on the Selection of Technical and Environmental Dangers* (Berkeley: University of California Press, 1982), 192.
18. Mary Douglas and Aaron Wildavsky, *Risk and Culture: An Essay on the Selection of Technical and Environmental Dangers* (Berkeley: University of California Press, 1982), 186.
19. Ezrahi, *The Descent of Icarus*, 24.
20. Ezrahi, *The Descent of Icarus*, 35.
21. Anthony Giddens, *The Consequences of Modernity* (Stanford: Stanford University Press, 1990), 34.
22. Kristin Shrader-Frechette, *Risk and Rationality: Philosophical Foundations for Populist Reforms* (Berkeley: University of California Press, 1991), 9.
23. Kristin Shrader-Frechette, *Risk and Rationality: Philosophical Foundations for Populist Reforms* (Berkeley: University of California Press, 1991), 12.
24. Kristin Shrader-Frechette, *Risk and Rationality: Philosophical Foundations for Populist Reforms* (Berkeley: University of California Press, 1991), 187.
25. Jasanoff, *The Fifth Branch*, 77. See 268, note 55 to chapter 4, regarding the link between the ideas of trans-science and regulatory science.
26. Jasanoff, *The Fifth Branch*, 77. See 268, note 55 to chapter 4, regarding the link between the ideas of trans-science and regulatory science.
27. Anne Chapman, *Democratizing Technology* (London: Earthscan, 2007), 85, 112.
28. This is Chapman's argument, and indeed at least in the United States some of our key environmental statutes are constructed to support this view. However, it also is true that risk assessment, at least as practiced within the US federal govern-

ment, does attempt to account for lack of data by applying uncertainty factors to risk estimates to account for database deficiencies. A discussion of how uncertainty factors are applied in risk assessment at US EPA may be found in the dose-response discussion at EPA's Risk Assessment Portal, http://www.epa.gov/risk/dose-response .htm.

29. Chapman, 92.

30. Chapman, 243.

31. Jasanoff, *The Fifth Branch*, 77.

32. Jasanoff, *The Fifth Branch*, 77.

33. Jasanoff, *The Fifth Branch*, 77.

34. For a non-technical explanation of the chemical risk assessment process, see US EPA, "Risk Assessment for Toxic Air Pollutants: A Citizen's Guide," originally published as EPA 450/3-90-024, March 1991, http://www.epa.gov/ttn/atw/3_90_024.html.

35. National Research Council, *Risk Assessment in the Federal Government: Managing the Process* (Washington, DC: National Academies Press, 1983). The NRC defines risk assessment as "the characterization of the potential adverse health effects of human exposures to environmental hazards" (18). The definition has since been expanded to include effects on wildlife and ecological systems.

36. See the web site of EPA's Risk Assessment Forum for a complete list of EPA risk assessment guidelines and related technical documents, http://cfpub.epa.gov/ncea/raf. For a thorough description of risk assessment concepts and how EPA uses risk assessment in decisions, see the web site for EPA's National Center for Environmental Assessment, www.epa.gov/risk.

37. Two EPA documents provide clear descriptions of the factors that EPA risk assessors and decision makers consider when evaluating the quality and completeness of data for use in assessments and decisions. See US EPA, *Summary of General Assessment Factors for Evaluating the Quality of Scientific and Technical Information* (EPA 100/B-03/001, June 2003), which may be found at http://www.epa.gov/osa/spc/pdfs/assess2.pdf; and *Risk Characterization Handbook* (EPA 100-B-00-002, December 2000), which may be found at http://www.epa.gov/osa/spc/pdfs/rchandbk.pdf.

38. US EPA, *Risk Characterization Handbook* (EPA 100-B-00-002, December 2000), http://www.epa.gov/osa/spc/pdfs/rchandbk.pdf, 11.

39. For a description of the Candidate Contaminant List (CCL) and the September 2009 CCL-3, see http://www.epa.gov/ogwdw000/ccl/index.html.

40. US Geological Survey, "Toxic Substances and Hydrology Program: Five-Year Plan, 2007-2011," 15 March 2007, 13, http://toxics.usgs.gov/pubs/pdfs/Toxics5-yrPlan2007-11_Final_3-15-07.pdf.

41. It is likely that no US agency makes a greater number of regulatory decisions than EPA, so a perusal of their chemical decision documents is a good place to look for how environmental decisions are characterized. A useful starting point is the regulatory action factsheets developed within EPA's Office of Pesticide Programs, which may be found at http://www.epa.gov/pesticides/factsheets/reg_fs.htm.

42. US EPA, "EPA's Endangerment Finding: Environmental and Welfare Effects," http://www.epa.gov/climatechange/endangerment.html.

43. Miller and Rose, *Governing the Present*, 32–33.

44. Scott, 24.

45. US EPA, *Exposure Factors Handbook* (Final Report) 1997, U.S. Environmental Protection Agency, Washington, DC, EPA/600/P-95/002F a-c, 1997.

46. Porter, 213.

47. Jasanoff, *States of Knowledge*, 38.

48. Ezrahi, 66.

49. Rose, "Governing By Numbers," 678.

50. Rose, "Governing By Numbers," 678, and Ezrahi, 33.

51. In particular, the hearings chaired by Senator James Delaney, which led to the insertion of the "Delaney Clause" into the Federal Food, Drug, and Cosmetic Act. The Delaney Clause said that no food additive would "be deemed safe if it is found to induce cancer when ingested by man or animal" and therefore a pesticide residue tolerance could not be set for that chemical "additive."See Christopher J. Bosso, *Pesticides and Politics: The Life Cycle of a Public Issue* (Pittsburgh: University of Pittsburgh Press, 1987), 97.

52. Ezrahi, 263.

53. Porter, 28.

54. Ezrahi, 17.

55. Giddens, 34.

56. Giddens, 35.

57. Jasanoff, *The Fifth Branch*, 206.

58. For illustrative exchanges among interest groups, see House Committee on Commerce, Subcommittee on Health and the Environment, *Food Quality Protection Act of 1995: Hearings before the Subcommittee on Health and Environment of the Committee on Commerce, House of Representatives*, One Hundred Fourth Congress, first session, on HR 1627, June 7 and June 29, 1995 (Washington, DC: US Government Printing Office, 1995).

59. Porter, 115, 200.

60. Porter, 200.

61. Paul R. Lockman, Joanna M. Koziara, Russell J. Mumper, David D. Allen, "Nanoparticle Surface Charges Alter Blood–Brain Barrier Integrity and Permeability," *Journal of Drug Targeting*, December 2004, vol. 12, no. 9–10: 635–641.

62. US EPA, *Nanomaterial Research Strategy*, June 2009, EPA document number 620/K-09/011, 1. The EPA strategy may be found at www.epa.gov/nanoscience. This is not to say that these four questions are the only ones to ask; indeed, it is my argument that this framework is inadequate.

63. For example, see OECD, "Guidance Manual for the Testing of Nanomaterials: OECD's Sponsorship Programme; First Revision," 2 June 2010, http://www.oecd.org/officialdocuments/displaydocumentpdf/.

64. Minimum Information for Nanomaterial Characterization Initiative, "The Parameters List," http://characterizationmatters.org/parameters.

65. Organization for Economic Cooperation and Development (OECD), "Draft Guidance on Sample Preparation and Dosimetry for the Safety Testing of Manufactured Nanomaterials," OECD document ENV/CHEM/NANO(2009)7/REV2, 16.

66. US EPA, "Nanomaterial Research Strategy," 11.

67. US EPA, "Nanomaterial Research Strategy," 12.

68. US EPA, *Ecological Risk Assessment Guidelines*, 1998, 34, http://cfpub.epa.gov/ncea/cfm/recordisplay.cfm?deid=12460.

69. Executive Office of the President of the United States, *National Nanotechnology Initiative Strategy for Nanotechnology-Related Environmental, Health, and Safety Research*, 21.

70. US EPA, "Risk Assessment Portal," www.epa.gov/risk.

71. US EPA, *A Summary of General Assessment Factors for Evaluating the Quality of Scientific and Technical Information*, EPA document number 100/B-03/001, *2003*. http://www.epa.gov/osa/spc/assess.htm.

72. US EPA, *EPA Nanotechnology White Paper* (2007), www.epa.gov/osa, 42.

73. US EPA, *Nanomaterials Research Strategy*, 11.

74. OECD, "Report Of An OECD Workshop On Exposure Assessment And Exposure Mitigation: Manufactured Nanomaterials," *Joint Meeting Of The Chemicals Committee And The Working Party On Chemicals, Pesticides And Biotechnology*, OECD document number ENV/JM/MONO(2009)18, 7 July 2009, 20. http://www.oecd.org/dataoecd/15/25/43290538.pdf.

75. US EPA, *Guidelines for Exposure Assessment*, 29 May 1992, Federal Register 57(104):22888-22938, EPA document number EPA/600/Z-92/001, 8. www.epa.gov/risk/guidance.htm.

76. Joseph Pitt, "When Is an Image Not an Image?" *Techné* 8:3, Spring 2005, 24–33.

77. Joseph Pitt, "When Is an Image Not an Image?" *Techné* 8:3, Spring 2005, 31.

78. Craig A. Poland, et al., "Carbon nanotubes introduced into the abdominal cavity of mice show asbestos-like pathogenicity in a pilot study," *Nature Nanotechnology* 3, 423–428 (2008), published online 20 May 2008. | doi:10.1038/nnano.2008.111. For a review of information developed on the link between carbon nanotubes properties and hazard, see Marie-Claude F. Jaurand, Annie Renier, and Julien Daubriac, "Mesothelioma: Do asbestos and carbon nanotubes pose the same health risk?" *Particle and Fibre Toxicology*, 2009, 6:16.

79. Environmental Defense and DuPont, *Nano Risk Framework, Nanomaterial Research Strategy*.

80. ED-DuPont, *Nano Risk Framework*, 15; EPA, *Nanomaterial Research Strategy*, iii.

81. Poland, et al., compared carbon nanotubes of different aspect ratios, and only found effects in the long, straight tubes. Poland, et al., 425.

82. Ezrahi, 17.

83. For the "twelve principles of green chemistry," see P. T. Anastas and J. C. Warner, *Green Chemistry: Theory and Practice* (Oxford University Press: New York, 1998), 30.

84. Hajer, 25.

3

Language and Nanotechnology

My coverage of regulatory science in chapter 2 demonstrates the critical role that the generation of hazard and exposure data has in establishing positions on the probability of a substance posing a risk to humans or other living things. This role is revealed in the statements that interest groups make about nanotechnology risks in debates dominated by regulatory science. In the United States, it does not appear possible to discuss environmental, health, and safety risks—perhaps generally but most certainly with regard to nanotechnology—without invoking the need for, or citing data from, regulatory science activities. This chapter's analysis supports my argument that language and practice within the nanotechnology risk discourse constrains debates on what is meant by environmental risk.

Discourse delimits the scope of inquiry—it determines what questions can and cannot be asked.[1] This chapter's discourse analysis shows the centrality of data in, and its role in defining, the nanotechnology risk debates. The ability to speak in terms of regulatory science data appears to be essential for having a voice in the debates. So far, nanotechnology has been treated within discourse much like industrial chemicals, in that regulatory science serves as the lens through which those engaged in the debates view environmental risks. It is a lens that has been focused narrowly on quantitative risk analysis, thus limiting and impoverishing nanotechnology risk discussions. My analysis of language points to both the problems of, and possibilities for overcoming, the limitations of this narrow focus.

The discourse analysis employed here recognizes the active role that language plays in social construction. While the term *discourse* can be defined many ways, its meaning for social science has been articulated by a number of scholars, many of whom use Foucault's work as a point of departure.

Dryzek defines discourse as "a shared way of apprehending the world. . . . Discourses construct meanings and relationships, helping to define common sense and legitimate knowledge."[2] Hajer considers discourse to be "a specific ensemble of ideas, concepts, and categorizations that are produced, reproduced, and transformed in a particular set of practices and through which meaning is given to physical and social realities."[3] What these definitions share is a view of discourses as active, involving constructions and practices—they are not merely words. Therefore, while my analysis of the nanotechnology risk discourse considers how language is used, it does so for the purpose of understanding how language reveals the practices within discourse that impact debates on risk.

Sara Mills makes the important point that discourses play themselves out within particular social contexts, and that discourses shape—and are shaped by—the social and cultural contexts within which they reside.[4] Useful for analysis of risk science language in the nanotechnology social context is the observation that discursive theories center around agreed-upon "story lines." These story lines, as described by Hajer, provide a narrative around which people with diverse perspectives on and orientations to an issue can gather and form coalitions to achieve an objective.[5] In the nanotechnology debates, the scientism story line draws actors together in the discussion and conduct of regulatory science. The statements analyzed in this chapter reveal the positioning of actors as they use the language of risk to articulate their views. This language fits into a larger science policy context within which debates are conducted on how to use science in environmental decision making.[6] Science's epistemologically privileged position in modern western society ensures that the statements provided here center around this "how" issue, and not *whether* science, in particular regulatory science, is the appropriate means for gaining understanding of the possible environmental implications of a new technology.

ANALYZING NANOTECHNOLOGY RISK LANGUAGE

Taking one at a time each of the three discursive categories described in chapter 1—risk society, technological progressivism, and administrative pragmatism—in this chapter I ask three questions about the use of language within statements made in the nanotechnology debates, adapted from Dryzek's 1997 analysis of environmental discourses.[7] The changes I have made to Dryzek's questions reflect my more-focused evaluation of language used to influence public policy in the United States, as well as my interest in furthering STS scholarship through study of the nanotechnology case. My three questions are:

- *What relationships are acknowledged?* Relationships may be reinforcing, such as with the relationship between quantification and legal defensibility. They also may be conflicting, as is the relationship between scientific uncertainty and legal defensibility.
- *What assumptions are made about those relationships, and how can STS scholarship help explain those assumptions?* Underlying every identified relationship are assumptions about that relationship that are expressed through discourse. For instance, in the relationship between quantification and legal defensibility, within administrative pragmatism it is assumed that experts will work within liberal democratic processes of public transparency and open participation to make their arguments. STS scholarship has something to say about why we have these assumptions about the need for experts, and how those assumptions fit within the relationship between science and the functioning of liberal democratic society.
- *What metaphors or other rhetorical devices are employed in the statements?* This question refers to how language is used, and includes syntax and use of key words that appear to have been selected to elicit a specific response or construct a particular meaning. The phrase "inherent promise of nanotechnology" is a good example of words joined together to convey an important set of messages within the technological progressivism discursive category.

In addressing these questions, I have drawn where appropriate from the STS literature to examine the ways in which the social construction of science and technology is manifest or described in language. In general, I have dedicated the most space to addressing the question, "What assumptions are made about those relationships?" because it is here where I have found the most-useful applications of existing STS scholarship to the new issues raised by the nanotechnology case.

Table 3.1 provides a summary of findings from my analysis of the statements I have located within the three discursive categories. While the table captures the main ideas described in this chapter, all 665 statements as a body of language provide a much richer and nuanced sense of the nanotechnology risk debate. To complement the sampling of statements provided in this chapter, the Appendix contains a series of charts illustrating my sorting of statements within discursive categories.

NANOTECHNOLOGY VOICES

The nine individuals whose statements are the data of this analysis have diverse professional backgrounds. Lynn Bergeson, an attorney, is a founder

Table 3.1. Discourse Analysis Summary

Characteristics *Discursive Category*	*Relationships*	*Assumptions*	*Language, Metaphors*
Risk Society	Technology-Society	Citizens have power to impact the direction and implications of technology.	Language concerning inherent qualities, uncertainty, dread, anxiety, power and powerlessness.
	Natural-Human	There may exist a distinction between an unnatural technology and a human society.	Statements that nanomaterials exist in nature, together with other statements touting human control over particles at atomic and molecular scales.
Technological Progressivism	Technological Progress-Societal Well-being	Human progress is tied to scientific and technological development.	Language of optimism, such as the promise and potential of nanotechnology.
	Public Perception-Technological Success		Language of manipulation, such as tools and pipelines.
	Technology-Governmentality Institutions		Language of movement and progress, such as marches and advances.
	Science/Knowledge-Nature		
Administrative Pragmatism	Articulators of Risk-Managers of Risk	Government is the primary authority, as well as the most responsible and accountable to society.	Language of authority: leadership, responsibility, accountability.
	Data-Decision Making	Scientific data can bring closure to risk debates.	Language of discipline and order: rules, standards, protocols.

and shareholder of Bergeson & Campbell, P. C., a Washington, DC, law firm concentrating on chemicals-related issues. She has been very active in engaging with federal agencies and the OECD on issues related to identifying and enhancing the environmental benefits of nanotechnology. Vicki Colvin is Kenneth S. Pitzer-Schlumberger professor of chemistry and professor of chemical and biomolecular engineering at Rice University. Colvin's research involves creating and studying nanoparticles for environmental applications such as water treatment, as well as investigating what properties of nanoparticles may be linked to toxicity. Richard Denison is a senior scientist at the Environmental Defense Fund. Denison has been active in nanotechnology and industrial chemicals, in particular regarding data generation, regulatory requirements, and risk assessment issues. Bart Gordon, former Democratic congressman from Tennessee, chaired the House Committee on Science and Technology. Gordon actively engaged in nanotechnology issues while in Congress, and convened hearings on nanotechnology and the environment after assuming the committee chairmanship in 2007. Until 2010, Andrew Maynard served as the science advisor to the Project on Emerging Nanotechnologies at the Woodrow Wilson Center for International Scholars. Currently director of the Risk Science Center at the University of Michigan, prior to joining the Wilson Center, Maynard was a physicist at the National Institute for Occupational Safety and Health. Terry Medley, an attorney, is global director of Corporate Regulatory Affairs at DuPont. Medley, who before going to DuPont was administrator of the Animal and Plant Health Inspection Service at the US Department of Agriculture, has been heavily engaged in nanotechnology environmental issues in his role of heading the Business and Industry Advisory Committee delegation to the OECD WPMN. Sean Murdock served as executive director of the NanoBusiness Alliance, a nanotechnology industry association. Murdock was an active lobbyist on Capitol Hill and participated frequently in nanotechnology conferences and workshops. Mihail Roco is senior advisor for nanotechnology at the National Science Foundation. Roco, a former professor of engineering, was one of the founders of the NNI, and plays a leading role in federal interagency coordination on nanotechnology issues. Until 2011, Clayton Teague served as director of the National Nanotechnology Coordination Office (NNCO), which is the secretariat to the Nanoscale Science, Engineering, and Technology Subcommittee, which coordinates NNI activities. Teague came to the NNCO after a thirty-year career as a physicist at the National Institute for Standards and Technology. Despite these individuals' diverse organizational backgrounds, with the exception of Murdock (who holds an MBA degree) all of them trained in the physical sciences, engineering, or law—fields that underpin regulatory science. It is now time to look at a sampling of the 665 statements made by these individuals, to illustrate how their language operates within discourse to advance a risk narrative dominated by scientism and regulatory science.

RISK SOCIETY

This quietest of the three discursive categories nevertheless provides insights into how science and technology are located in policy-related discussions.[8] Woven throughout statements made in the spirit of Beck's reflexive modernization are implications for power relations and the place of science in society.

What Relationships Are Acknowledged?

How science and technology, viewed as a single entity and perceived as a force unto itself, interacts both with and within society is an important relationship within the risk society discursive category. This idea joins with another dominant relationship: distinctions and interactions between the natural and the human. Two views from the STS literature, one by Donna Haraway and another by Bruno Latour, point to philosophical issues that are highly relevant to considering how the nanotechnology debates describe the relationship between what is natural and what is of human creation. Both Haraway and Latour advance approaches for blurring what they see as an unnatural divide between the natural and the human. However, Haraway retains nature as a distinct, singular entity, open for reinterpretation but not capture by culture. Latour does not admit the existence of nature without culture, or vice versa, and therefore advances a resolution that acknowledges their symbiotic and co-produced relationship. A statement from Andrew Maynard in response to the question, "But is it realistic that ordinary citizens will be able to influence the direction of nanotechnology?" begins to set up this relationship within nanotechnology risk language from a risk society perspective.

> It's very clear that they can influence which technologies succeed and which ones fail. If people decide they don't want nanoparticles in their products, it is entirely conceivable that you'll have a movement against that technology, which will knock that technology on the head. There is also a fear about risks, that there's not enough information out there. A movement like that will also affect other nanotechnologies. People don't differentiate between different types of nanotechnology. You could have people making decisions on nanoparticles in a sunscreen, but then also use that approach to how you use carbon nanotubes. That really would be very damaging to technology as a whole, because people making decisions which aren't based on the science,

but on the gut feeling of what they like or don't like, and you're not going to change how those people make decisions fundamentally.[9]

With Maynard talking of people taking up against a nanotechnology application to "knock that technology on the head," one gets the sense of technology being an entity unto itself, separate from the people who would do violence to it. As in so many of the statements in this study, the overarching theme of scientism is prominent here, with the speaker implying that decisions not based on science will do undeserved broad damage to nanotechnology as a whole. Yet Maynard acknowledges that "gut feeling" actions are part and parcel of the risk society, as people use what tools are available to them to push back against technologies they view as harmful. What makes this a true risk society statement is its power implications. Maynard notes not only that people will take action against nanotechnology, but that they *can* take action, with or without scientific information. This potential for de-centering of science clearly worries him.

A different perspective on what is human versus what is natural is given by Environmental Defense Fund's Richard Denison, who stated in a congressional hearing that:

The precise and highly homogeneous composition of most engineered nanomaterials, and their intended use in specific applications, could well lead to exposures of a wholly different nature and magnitude than those associated with natural or incidental nanomaterials.[10]

Here Denison suggests that nanoparticles made through human technology are more dangerous than nanoparticles found in nature. There is something more menacing in the particles because of their technological origins, and because of their "intended use" in products introduced into society. Power in this statement resides in technology, which is able to produce "engineered" particles with a precision and uniformity unmatched by nature. The view that the fabrication and use of such particles "could well lead to exposures of a wholly different nature and magnitude" suggests a risk society anxiety that what is novel and technological likely will bring new, yet-to-be-identified sets of problems into human life.

The third and final statement presented in this section comes from chemist and nanotechnologist Vicki Colvin who, when talking about applying environmental sustainability approaches to developing nanoparticles, states:

There are many reasons to be concerned about nanomaterials. The good news is that nanomaterials exist in nature. . . . We now don't ask the question, "Are engineered nanomaterials dangerous?" We ask, "How can we make them safely?"[11]

This is an overlap statement. It falls within the risk society category because its starting point is noting the "many reasons" why nanoparticles should be of concern. Yet it also expresses confidence, in the spirit of both technological progressivism and scientism, that scientists and engineers can design-out these inherent dangers. Most relevant to the relationships being studied here, we see a linking of engineered particles to nature in order to argue for their potential to be safe. However, the statement suggests that such particles can only be safe if technologists intervene to imbue them with properties that make them compatible with nature.

What Assumptions Are Made about Those Relationships, and How Can STS Scholarship Help Explain Those Assumptions?

The animation of technology, together with the notion that when coupled with human intent technology begets unnatural objects that impose new threats upon society, raises an important assumption: that there exist distinctions between natural and human creations. Haraway develops the notion of boundary breakdowns between technologies and people. She identifies three such boundaries—between human and animal, between animal-human and machine, and between physical and nonphysical—that technology, in its interactions with people, tends to blur.[12] Because Haraway speaks of cultural groups as having been produced through the exercise of power and dominance within society,[13] it is useful to first define *society* as the specific context within which a particular culture thrives: *culture* is the organism grown in the societal Petri dish. By identifying humans and nature as "inhomogeneous," Haraway makes clear that *nature* is that which is apart from humanity; that is, *humanity* as we through culture in society have defined ourselves.[14] However, that does not make nature a clearly identifiable entity or, for that matter, concept. By employing scare quotes around "nature" and identifying it as a "protean trickster," Haraway gives nature an elusive identity that shifts in form as we vary our relation with it, and eludes our attempts to give it definition.[15]

For Bruno Latour, there is a Nature and there is a Society, but both are produced through human-nonhuman interaction.[16] They only exist within the context of construction. In Latour's theory, *culture* does not exist as a legitimate concept. The notion of culture was created by the West to distinguish itself from "others" who are not the holders of science-derived knowledge.[17] The others have culture—the West does not have a culture because we have mobilized nature through science thereby creating divisions that have allowed anthropology to develop "asymmetrically."[18] We created culture by dividing nature from society.[19] Since such a division is illegitimate, culture is not a legitimate concept. Latour describes two

Great Divides: a modernist divide between nature and society, and an anthropology-constructed "premodern" divide between the West and the others who do not hold nature separate from their societies. Anthropology is asymmetrical because it only looks at "cultures" (i.e., nonmoderns) and not back at modernity.[20]

Latour sees paradox as the principal implication of the constructed nature-society divide. Modernists set nature apart by identifying nonhuman Laws of Nature, yet they create these laws within their laboratories. The experiments are clearly constructed, yet the modernists maintain that their experimental results are not. Can nature exist unconstructed yet be constructed in the laboratory? Latour asks.[21] This paradox is manifest in *hybrids*: outcomes of the intersection of the social and the natural. As these hybrids are acknowledged and become prominent, the modernist model of purification becomes untenable. Latour discusses hybrids in terms of generalities, but they are easy to identify. When science is used in policy making, they are particularly obvious. For instance, the socially constructed area of climate science gains its legitimacy from observations of changes in nature (as something apart from the observer), yet the entire framework for measuring and interpreting the results of those observations is imbued with social interests associated with changes in temperature.

The Haraway and Latour perspectives on the nature-culture dynamic raise interesting issues, specifically for the three risk society statements discussed in this section as well as generally for discussions related to emerging technologies such as nanotechnology. A strength of Haraway's theory is its identification of the role of power in the social construction of the nature-culture relationship. As an emerging technology, nanotechnology may prove to be destabilizing both scientifically and culturally (*both,* of course, because they are jointly socially constructed). This potential for destabilization already appears to be manifesting itself in atypical interactions between interest groups; for instance, a partnership between Environmental Defense Fund and DuPont, and unusually cooperative engagement between nations with competitive commercial interests in nanotechnology.

Latour's concept of hybrids is useful for examining nanotechnology risk statements. In congressional hearing rooms and at scientific meetings, where language emphasizes modernist purification, hybrids nevertheless abound. In the nanotechnology case, claims will be made about "quantum effects" and their unique relevance, or lack thereof, to potential human or other biological systems, as measured through laboratory animal studies or computer models. But within these so-called unnatural, engineered nanoparticles, what could be more natural than quantum effects? The risk society language appears to represent inchoate, conflicted thinking on the source and location of nanotechnology risks in modernity.

What Metaphors or Other Rhetorical Devices Are Employed in the Statements?

Within the statements I have located in the risk society category, few metaphors are employed. Instead, adjectives are used to infuse nouns—such as "particle" or "the future"—with feelings of anxiety and uncertainty. Terms describing inherent risks of nanomaterials seem to be the most commonly repeated in this category, followed by those articulating an uncertain and uneasy future for nanotechnology. I will begin with inherency. In the following statement, Richard Denison testifies at a 2007 congressional hearing:

> There is essentially no tracking of the production and use of nanomaterials. This is another reason why it is so important to gain an understanding of the *inherent hazard* of a material, which is relevant no matter how it may be used or encounter people or the environment. [emphasis added][22]

He states in another hearing two years earlier:

> Primary concerns about nanomaterials' health and safety impacts arise both from consideration of the *inherent nature* and novel properties of at least certain nanomaterials, and from surprising results seen in many of the relatively small number of nanotoxicity studies conducted to date. [emphasis added][23]

Vicki Colvin uses a bit more catchy language to provide another take on the inherency issue:

> Every new technology brings with it a set of concerns that, if handled poorly, can turn "wow" into "yuk" and ultimately into bankrupt as the genetically modified foods industry discovered.[24]

Here is a final example, from Andrew Maynard, of a statement that goes to the inherency issue, in response to the following interview question: "Now, industry patents nanoparticles because of their unique properties, but fights regulation on the grounds they're just smaller versions of common materials. How do you reconcile this?"

> It's very, very difficult to reconcile. One of the issues I've always had is that it's very easy to be narrow-sighted. As you said, it doesn't jive at all, that they can say on the one hand, these things behave in unique and unusual ways, and on the other hand, when you're looking at risk, they're the same as everything else. We've got to be a little more sophisticated in our thinking and realize that if something is going to behave in unusual ways, there's got to be the possibility of unusual risk associated with it.[25]

Whether a statement attributes to nanoparticles an "inherent hazard," "inherent nature," a "wow-to-yuk" potential, or observes that calling the same particle both unique and common "doesn't jive [*sic*]," an important message conveyed by these statements is that the nanoparticle carries within itself a source of dread or anxiety. Language within the risk society category exposes nanotechnology's negative image: a tonal inversion of the technologist's colorful computer-generated picture of ordered, engineered particles, where through the scrutiny of reflexive modernization the brightly colored representations of atoms are rhetorically darkened with the language of inherent danger, foreshadowing an assumed future risk when the particles will escape from their intended uses and find their way into the environment.

A number of statements speak to nanotechnology's uncertain future. I have chosen three to highlight here. While they all approach both the future and uncertainty from different angles, they have in common the idea that humans will have to intervene to *do things to technologies* to make them acceptable. This aspect of reflexive modernization—that unexamined technologies are dangerous technologies, but that such examination can lead to socially and environmentally acceptable fixes—is a form of language that, based on this study, appears to dominate the application of risk society language within US environmental policy discussions.

The following statement by Andrew Maynard makes a somewhat hopeful, but nevertheless uncertain, statement about nanotechnology's future:

> Current understanding of nanomaterial risks has more holes than Swiss cheese. . . . If we are going to get a good handle on working safely with engineered nanomaterials and other products based on nanotechnology, these *holes* will eventually need to be filled. . . . Even though we are facing a nanotechnology safety future that is complex and riddled with holes, we do have some tricks at our disposal for helping to ensure the *safer* handling of nanomaterials. . . . On the other hand, it's probably not a good idea to be complacent—old tricks may work with new technologies, but probably only up to a point. [emphases in original][26]

The idea of holes being in a technology's future, together with the message that such holes can be filled through human intervention, is a repudiation of the technological progressivist notion that technologies progress down predetermined paths and must be managed only after their introduction into society. Rather, Maynard is arguing with risk society language that technologies come with inherent problems, that left to their own devices—or, perhaps more accurately, to unbridled technology boosterism—will cause them to be unsafe. Up-front hole filling is needed by reflexively evaluating and then, perhaps even more crucially, taking action on the technology to address its risks.

Denison puts uncertainty, and the perils of inaction, into historical perspective:

> History demonstrates that *embracing a technology without a careful assessment and control of its risks* can be extremely costly from both human and financial perspectives. The failure to sufficiently consider the adverse effects of using lead in paint, plumbing, and gasoline has resulted in widespread health problems that continue to this day, not to mention extremely high remediation costs. Asbestos is another example. . . . [emphasis added][27]

Again we hear a clear call for anticipatory risk assessment of technology, and a repudiation of technological progressivism's unconditional "embracing" of technologies. Like Maynard, Denison believes that inserting risk considerations into emerging technologies—taking "control"—is both possible and necessary. He also invokes risk society reflexivity by looking back at modernity's past mistakes.

We come back to Maynard, who in a blog puts people at the center of nanotechnology's future:

> Without that ingrained culture of putting others first, I wonder whether there is a danger of nanotechnology risk research being driven more by political expediency and the promise of economic gain, and less by the need to protect people. . . . We need to work out how to get *people* back at the center of the nano-risk enterprise. . . . getting nanotechnology "right" will be a hollow achievement if we end up neglecting the very people who will make its success possible. Let's hope we don't.[28]

In this short statement, he has identified three important power relationships. The first is an explicit identification of economic and political drivers of technology as potentially damaging forces. Rarely in the statements examined in this analysis are political and economic power directly identified as negative; typically, they are characterized as important forces to be marshaled to advance nanotechnology. Maynard softens the attack with the use of "I wonder," but the intent is clear. The second power relationship is the one we have already discussed in the previous two statements: the power of human intervention to alter nanotechnology to make it less risky. The third is the power of the public to determine the success of nanotechnology. The implication here is that the public will be the deciding factor in whether nanotechnology becomes successful, irrespective of whether risks are proactively addressed. (Just as the American public, by accepting genetically modified [GM] foods, has made GM crops tremendously successful in the United States.) However, Maynard introduces a moral imperative by stating that the public acceptance of nanotechnology will be a "hollow victory" if those with the wherewithal to address risk issues fail to do so.

This analysis reveals important attributes of Beck's concept of the risk society employed within the nanotechnology risk context. First, statements articulate nanotechnology as a force in and of itself, with its own inherent risky attributes. However, the statements also indicate confusion about whether it is a force of nature or of human making, and we have explored whether such a distinction is even possible. Added to this confusion is the uncertainty over nanotechnology's future: Will a reflexive approach to its introduction lead to it becoming a "successful" technology? Unanswered questions include what such success would mean, and for whom.

TECHNOLOGICAL PROGRESSIVISM

The technological progressivism described by Kleinman exhibits great certainty that society can have little if any influence over the path of technological development.[29] However, in the nanotechnology statements I find a bit more equivocation and nuance than might be implied by Kleinman's description of this discursive category. The rounding of technological progressivism's hard edges is not due to any diminished faith in the value of technology for improving society, but rather because of suggestions that nanotechnology's path, while clearly marked out, is not yet laid in stone. Here I examine such nuances, mindful that despite some equivocation, the general lack of reflexivity regarding technology's place in modernity merits locating these statements within technological progressivism.

What Relationships Are Acknowledged?

Four principal relationships emerge from the technological progressivism statements: (1) technological progress and societal well-being as both interdependent and inevitable, (2) a negative association between public perceptions of risk and the speed of technology development, (3) the relationship between technology and governmentality institutions such as regulatory frameworks and economic systems, and (4) interactions between science/knowledge and nature. For each of these relationships, I have selected one statement to illustrate the various ways in which technological progressivism characterizes nanotechnology risks.

The following statement from the NSF's Mihail Roco speaks to nanotechnology's inevitability and its potential for societal benefit.

> Developing knowledge at the nanoscale is a natural trend in science and engineering. This may prepare us to address unexpected risks of human activity such as encountering unknown viruses and bacteria. . . . Nanotechnology has the long-term potential to bring revolutionary changes in society and harmonize international efforts towards a higher purpose than just advancing a single

> field of science and technology, or a single geographical region. A global strategy guided by broad societal goals of mutual interest is envisioned.[30]

Here we see that nanotechnology's "natural trend" of development will improve society. The inclusion of "towards a higher purpose" is interesting. Roco is suggesting that extending human control over nature at the nanoscale will bring scientific disciplines and technological efforts together for global good, including using our new power of control at the atomic and molecular scales to fight new diseases ("unknown viruses and bacteria"). According to Sarewitz, Roco's statement is reflective of what Sarewitz calls "the myth of infinite benefit," in which the enhancements in the quality of human life are directly linked to increased resources devoted to scientific research.[31]

Within technological progressivism, anti-technology sentiments run counter to this natural trend and higher purpose, and therefore are seen as negative social forces. A number of statements identified as important the need to persuade the public of nanotechnology's benefits. The statement below by Vicki Colvin is an excellent example.

> The perception that nanotechnology will cause environmental devastation or human disease could itself turn the dream of a trillion-dollar industry into a nightmare of public backlash. This negative response is possible even if the environmental and health threats never materialize. To nanotechnology researchers like myself, that prospect is all too real, and just as frightening as anything a sci-fi writer can imagine. . . . If we fail to answer these questions early, public acceptance of nanotechnologies could be in jeopardy and the entire industry derailed.[32]

The "nightmare" for Colvin quoted above is not that nanomaterials will turn out to be risky, but that a public backlash will jeopardize the development of nanotechnology. That the nanotechnology industry could be "derailed" implies that the track has been laid and the destination determined. Colvin finds "frightening" the idea that risk perception could knock the technology train off the rails. Here also we find articulation of a power relation that shows up in a number of statements, and was touched on in the last statement used in the risk society analysis: that the public has the power to impede technological progress. However, whereas in risk society language such power is a source of optimism and opportunity for putting technology on a socially beneficial path, for the technological progressivist the idea of the public wielding this sort of power is of considerable concern.

Because technological progressivism expects society's controlling institutions to provide motive power behind the development and diffusion of technology, within technological progressivism a close relationship exists between technology and social institutions that can be identified in the *discursive character of governmentality*. That is, as described by Miller and Rose, the language used by such institutions is rendered into particular forms shaped through political discourse.[33] For nanotechnology within technological progressivism, that means social institutions are described in terms of their ability to advance, both through promotion and the removal of barriers, the development and deployment of nanotechnology. The statement below by the NNI's Clayton Teague illustrates this relationship.

> The view that we have is that the current regulatory system that is in place from EPA, from FDA, from CPSC, from OSHA are such that they are appropriate for handling any of these new materials. They have each taken special attention to any of the new nanomaterials. Each one of the agencies that I have mentioned have formed task forces within their agencies to pay special attention to nanomaterials and to consider how their regulatory authorities would apply to such materials. The other aspect of this issue is that it is the responsibility of the manufacturers to ensure that any materials, any products which come on the market, are safe. The regulatory agencies are there to make sure that that is true, but it is the manufacturers' responsibility to ensure that products, materials, devices are safe when they come to the marketplace, and we feel confident, I feel confident, that that is the case with respect to nanomaterials.[34]

Here Teague has outlined responsibilities for government and industry for ensuring nanomaterial safety. We note that in this statement government agencies are positioned "for handling any of these new materials." Industry, in turn, is to ensure their safety "when they come to the marketplace." The relationship of government and industry to technology is one of shepherding the technology forward (in a safe way), and not of questioning its utility, desirability, or even its development path. Shared responsibility implies a common interest in an outcome, and here the outcome is the safe, successful introduction of nanotechnology into society.

The final relationship identified within this discursive category is interactions between science/knowledge and nature. Technological progressivism describes this relationship generally in terms of using science and technology to exercise control over nature. Control at the nanoscale is viewed as a step toward controlling atoms and molecules, the fundamental building blocks of matter—the ultimate human control over all else. Another statement by Vicki Colvin articulates a common sentiment within technological progressivism.

> One of the real powers of nanotechnology is that we have the capability to engineer in a controllable way the property of matter at this nanoscale. That has really important implications, in terms of what was mentioned relative to, say, the buckyballs. Because we have this capability to control things and engineer things at the nanoscale, we can study them. We know how to, now,

> treat the buckyballs so that they will not be detrimental to human health or the environment. We know how to functionalize the surface of these small particles in such a way to make them more benign to both public health and to the environment. So, we happily are at an early stage, where we not only can study these before the widespread application of the technology and production of large volumes of products, we can understand their behavior, and we have sufficient control to where we can engineer them to be what we would like the properties to be, and hopefully, avoid the negative properties, in terms of adverse impact upon the environment, or upon public health.[35]

Here power over nature is cast in terms of risk avoidance. By controlling the properties of particles, we can engineer risk out of nanomaterials. Colvin recognizes that control of matter at the nanoscale involves risk, in terms of making potentially harmful particles, but expresses with great confidence that we can both understand the potentially harmful properties and control for them in our manipulation of matter. Such confidence is a hallmark of the language of technological progressivism.

What Assumptions Are Made about Those Relationships, and How Can STS Scholarship Help Explain Those Assumptions?

All four relationships described in this analysis of technological progressivism are underpinned by one large assumption: that some identifiable (if not objectifiable) measure of human progress is tied to the advance of science and technology. In large part, human progress is *defined* through scientific and technological progress. Because this assumption is central to the identity of this discursive category, it would be useful to consider it from the perspective of the STS literature.

A starting point for exploring this assumed link between S&T and human progress is the work of Émile Durkheim introduced in chapter 1. In Durkheim we see perhaps the first deliberate articulation of a relation between scientific knowledge and society. Central to Durkheim's argument is an understanding of the *sacred*. According to Durkheim, humans identify things as sacred to reconcile the reality of their everyday (profane) existence with the promise of higher-order (sacred) possibilities they imagine for themselves.[36] As thinking, conscious beings, we imagine a greatness beyond ourselves. Most important for Durkheim is that individuals are able to construct representations of the sacred only because they are able to conceptualize, and concepts come from society, not from individuals.[37]Durkheim makes the case that, as with other aspects of society, science developed out of religion. Our faith in science is analogous to our faith in religion, and the nature of that faith will change just as human notions of religion have changed over time.[38] As the discussion in chapter 2 suggests, the faith that decision makers have in regulatory science's ability to support their actions is consistent with Durkheim's assertion of science having assumed some cultural functions previously carried out by religion.

Whereas in *The Elementary Forms of Religious Life* Durkheim examined society's relation to science as an institution, early writings by Robert Merton center around the scientific information producers' relation with themselves and society at large. Merton's fundamental position is that science has a particular ethos and certain norms, and that the scientific ethos and norms can be identified and examined through the study of scientists' writings and behaviors.[39] The four norms—universalism, communism, disinterestedness, and organized skepticism—for Merton describe the ideal relation among scientists.

Merton realizes, like Durkheim, that scientists are situated with a larger societal context. From Merton's view, the relation between scientists, the knowledge they produce, and society at large is inherently unstable. The scientific community must continually reinforce the normative pillars of its ethos if it is to withstand seismic tremors of social change. War and other civil unrest can undermine societal belief in universal truth, jeopardizing the acceptance of scientific knowledge claims.[40] While Durkheim's science has a prominent edifice (temple?) in the town square, Merton's science comes across more like a sect whose members stand on street corners handing out pamphlets to passersby. The tension between Durkheim's idea of science as sacred, and Merton's scientific ethos and norms, is a useful contrast when evaluating truth statements from science as an institution of governmentality, and scientists both as interest groups and actors in liberal democratic society.[41]

Simon Schaffer and Steven Shapin argue that the English power elite and intelligencia's adoption of the experimental method arose from the political environment of the Restoration, in particular from the desire to maintain social order in post-civil England. In *Leviathan and the Air Pump*, large social forces shape the direction of both the content and manner of scientific inquiry.[42] Shapin to some extent builds on this argument in a 2008 book that examines the development of science as profession, primarily in twentieth-century America. In this work, Shapin describes how the disruptive societal transformations of the past century—which he labels the "normative uncertainties of late modernity"—have shaped who scientists are, where they have become employed, and what knowledge systems they have pursued.[43] Shapin concludes that within an increasingly uncertain and unstable societal context, success in science has depended increasingly on individual characteristics of the scientists themselves: their values, charisma, and recognition within certain influential segments of society.[44]

Shapin's conclusions are in some respects at odds with both Durkheim and Merton. In the world Shapin describes, capitalistic liberal democracy is the religion and the scientist at most is something between an acolyte and a high priest. Most often, though, the scientist is a member of the congregation, performing important duties such as ushering in new findings and tools that can be applied to advancing societies' objectives, small and large. There is somewhat less tension between Shapin and Merton. Virtue matters in Shapin's view of science in late modernity, as it did for Merton decades prior. Yet there is an important and useful distinction. Whereas for Merton the scientific ethos engendered norms internal to the institution of science, Shapin sees scientific virtue as embodying the best characteristics of the liberal democratic citizen: integrity, ambition, and responsible individualism.[45]

Durkheim is right that our collective representation of what we as society view as knowledge is firmly tied to a vision of science as human elevation above the profane toward the sacred.[46] In science, society feels larger than it would otherwise. As part of society, ecology and environmentalism have joined themselves to that confederation of hypothesis-driven approaches we call science—we believe that we need to *know* Nature to save her. And technology is a means of such salvation. For technological progressivists, this is as true for nanotechnology as it is for other environmental risk issues. Technological progressivism voices clear and confident support for the Enlightenment project. It implicitly rejects Postmodernism, mostly by simply ignoring it. It is, following Durkheim, a way of talking about the world in a voice more clearly American than either risk society or administrative pragmatism, and is a strong voice in nanotechnology risk debates.

What Metaphors or Other Rhetorical Devices Are Employed in the Statements?

Words that convey optimism are generously used by technological progressivists. *Promise* and *potential* are examples. Mechanical metaphors are also common. There are *tools* to be employed and *pipelines* to carry materials from the laboratory to the factory to use. Finally, within technological progressivism is the language of movement, as technology's progress *marches* and *advances* in its development and use.

The following statement by Lynn Bergeson illustrates how describing the promise of nanotechnology can be linked both with commercial success and environmental protection.

> There are many promising agricultural applications of nanotechnologies. Enhanced use of smart systems could also diminish run off and avert unwanted movement of pesticides. These are only a few of the innovations nanotechnologies offer in the food and agriculture areas. New agricultural/antimicrobial products and application techniques are likely to revolutionize these markets, and there are many commercial opportunities to promote sustainable agricultural and pollution prevention through nanotechnologies. Industry stakehold-

ers and others must engage with EPA and the United States Department of Agriculture early, openly, and regularly to ensure nanotechnologies fulfill their promise as pollution prevention and sustainable agricultural tools.[47]

Here we see words like "promise" and "opportunities" to describe the application of nanotechnology to pesticide uses. Such promise is seen as revolutionary, and carrying the potential to move pesticides into a sustainable, post-risk future of sustainable use. Within technological progressivism, new technologies fix risk problems (created by old technologies) and promote environmental protection.

Regarding metaphors related to mechanics and movement, we have a statement from Vicki Colvin that uses both types of metaphors in describing how to create environmentally benign nanomaterials.

> Nanotechnologies offer new approaches to treating cancer and cleaning water . . . but fewer of these transformative technologies will make it into commerce if the technology transfer pipeline becomes clogged with concerns about nanoproduct safety. . . . The only sure fix is high quality and intelligently packaged risk-related information. . . . We have to explore the latest tools of twenty-first century biology to move from observations of nanoparticles hazards to a paradigm that seeks to predict these hazards before a material is even created. . . . Perhaps the first step along the path will be programs that encourage researchers to adopt a common set of tools for risk research.[48]

There is much to consider in this short statement. Here we see a danger of "transformative technologies" being jeopardized if the road to market is "clogged with concerns" about the safety of these technologies. Only "high quality and intelligently packaged risk-related information" can keep the pipeline free of such clogs. The idea that scientific information on risks must not only be of high quality, but also be well packaged, suggests in this context a strong belief in science's role in delivering solutions to decision makers. An implication is that if decision makers appropriately apply the information science supplies, concerns will disappear and nanotechnology's societal benefits will be fully realized. Interesting also is the notion of risk concerns as barriers to technology development. A barrier by definition is exogenous to that which attempts to avoid it. The statement therefore suggests that the voicing of societal concerns may not be a legitimate part of technology development, but rather is an obstacle to be overcome on the road to commercialization. A statement by Clayton Teague uses similar language.

> The road to effectively safeguarding public health is dynamic and requires attention to alternative paths for obtaining the necessary knowledge and understanding. An ongoing evaluation of the materials being considered for use in products and new research findings must continue to inform and guide our ongoing strategic planning efforts.[49]

We see a tension between Teague's inclination toward public safety and his tendency toward caution and careful planning. The notion that addressing risks is a process also comes through, with words such as "road," "path," "ongoing evaluation," and "ongoing strategic planning efforts" dominating the statement. What we also see is a public health goal arrived at via paths to knowledge. Ensuring public health with respect to nanotechnology thus is described as a destination, rather than as a current state to be maintained. Risk is not mentioned in the quote. This omission is common in technological progressivism language, where words such as "impacts" or "implications" are often used in place of "risk."[50]

In this statement, the implications for risk's influence on power relations are subtle, but significant. Implied is that the federal bureaucracy is in no position at this time to ensure life or even well-being in the face of potential nanotechnology risks. New scientific and technological roads must be built, or at least old ones repaved, to arrive at the knowledge that the government believes is needed to safeguard health. While this position on its face appears weak, it in fact holds considerable power. The power lies in the road building. The statement supports the position that the federal agencies—through academic grants such as those from the NSF, EPA, and the National Institutes of Health, as well as through research in federal laboratories—are the avenues and the vehicles for securing the scientific knowledge required to identify and address potential adverse consequences from nanotechnology, and by doing so clear the way for nanotechnology's introduction into society. As noted by Kleinman and Sarewitz, this position reinforces the longstanding role of the federal government in funding and conducting scientific research, going back in US history at least to Vannevar Bush if not further, and supports the power structure that has built up over time to affirm and maintain that role.[51]

ADMINISTRATIVE PRAGMATISM

Problem solving by experts, within a context of liberal democratic freedom, transparency, and public participation, is the set of ideas around which the language of administrative pragmatism revolves. Here we see most prominently the language of technocracy, democracy, and institutions. Each discursive category has its equivocations, since issues of environmental protection and risk are fraught with uncertainties. Yet among the three categories, administrative pragmatism seems the most confident in asserting that America's best hope in getting nanotechnology right resides in the institutions of liberal democracy.

What Relationships Are Acknowledged?

The relationship at the center of administrative pragmatism is between those who articulate nanotechnology risk positions and those who are charged to do something about risks. This relationship is central to describing how environmental issues are discussed and addressed. The following statement by Richard Denison illustrates the relationship.

> The conflict also appears to be manifesting itself at the individual agency level. Some NNI [National Nanotechnology Initiative] agencies, including FDA and EPA, are themselves charged with both promoting and regulating nanotechnology applications, sometimes even within the same office. In addition, all agency proposals pertaining to addressing nanotechnology's potential risks must now be vetted through a White House nanotechnology policy group. These factors may be responsible in part for the growing disconnect between, on the one hand, the recognition by agencies of the magnitude of an urgent need to address the risk question, and on the other hand, the tepid response of those same agencies in terms of actions to be taken.[52]

Key words and phrases in this statement include "conflict," "disconnect," and "urgent need to address the risk question." Conflicts and disconnects in this context apparently refer to what Denison sees as divided and incongruent institutional mandates and responsibilities. A result is inconsistent actions by those institutions. What he does not see as unclear is the "urgent" risk situation. The statement takes as a given that there are risks to be addressed, and that the responsibility for addressing those risks resides with the federal government.

This statement is given from the standpoint of an outsider looking in at federal regulators. Denison seems to be reinforcing the border-center distinction articulated by Douglas and Wildavsky. The center supports the status quo and thus populates the institutions that have an interest in making the future an extension of the present. The border stands in opposition to the institutions and envisions, because of its risk concerns, a future of disruption and uncertainty.[53] Here Denison appears to assume the role of representative for Agamben's *homo sacer,* sacred man who "cannot be sacrificed yet may be killed."[54] The institutions of government are duty bound to protect society from nanotechnology risks, which he sees as urgent (which could be interpreted as *life threatening*), yet these risks are ambivalently addressed by government. If "letting die" is the same as killing, then Denison's perspective could suggest that, with respect to nanotechnology, all the unprotected indeed are *homo sacer.*

The second essential relationship described within administrative pragmatism is between data and decision making. It is assumed that decisions must be data driven and that better data, and more of it, will lead to sound decisions on nanotechnology risks. This relationship is implied in the following statement by Lynn Bergeson.

> It does not appear to be useful to call for the submission of any and all information on the effect of the use of nanotechnology in any regulated product. For confidentiality reasons, manufacturers are unlikely to submit the most useful information, and will likely submit general data. Many companies may submit the same information. FDA will have difficulty reviewing and processing the information that is submitted, and it would be challenging to identify the types of product for which information should be submitted. Much of the information would not be relevant to any particular product FDA is required to review, and there is no telling whether the information submitted could be used to prepare the guidance documents recommended in the Task Force Report. *A more orderly submission of targeted materials on the basis of identified and specific issues would likely lead to the generation of more useful information.* [emphasis added][55]

The *information* this statement references is scientific data; specifically, it is data generated under the regulatory science construct. Here it is useful to note the centrality of order and specificity of purpose in the generation and use of data within the context of administrative pragmatism and the scientism/regulatory science story line.

A good starting point is the italicized sentence concerning "orderly submission" of "targeted materials." In this case, Bergeson argues for product-specific data, in effect advocating for a case-by-case evaluation of specific nanomaterial uses. The idea of generating and using well-ordered data to inform a narrow set of issues is a hallmark of both regulatory science and administrative pragmatism. Not only are the data generated along well-defined disciplinary lines, but expert decision making also is encouraged. Here, then, we see manifest in the nanotechnology case the ordering of nature described by Porter and Ezrahi as introduced in chapter 1.

Bergeson's statement indicates that it will be "useful" to generate and provide data to regulators on nanomaterial- and product-specific risk attributes. This raises a number of assumptions regarding the place of regulatory science information in environmental decision making.

What Assumptions Are Made about Those Relationships, and How Can STS Scholarship Help Explain Those Assumptions?

Administrative pragmatism assumes government to be the principal identifier, describer, and resolver of risk issues. This view raises a problem for risk and power relations in liberal democracy that once again takes us back to Beck and the risk society. According to Beck, what is different about modern human health and ecological threats, such as those from nanotechnology, is that the political conflict they create is not divided between classes, since threats to health or well-being, Beck argues, do not necessarily acknowledge class distinctions.[56] This situation creates a "politics of knowledge," in which those who understand the nature of modernity's risks coalesce in opposition to the producers of risks. Therefore, as boundary actors the environmental NGOs may see their role as information disseminators in the politics of knowledge, since they have little faith that the center, EPA and FDA in this case, will take the necessary actions to avoid or reduce nanotechnology risks. However, what Beck does not address—and what the NGOs are keenly aware of—is that even if nanotechnology risks have the potential to touch all people irrespective of class, opportunities for risk avoidance are stacked in favor of those with greater social means and political power. Take the example of labeling nanotechnology products, something supported by NGOs. Those people with the highest education and greatest incomes are the ones who are most likely to pay attention to product labeling and to purchase products with green attributes. They, in all likelihood, will be the greatest beneficiaries of any future regulatory actions, such as labeling requirements, taken on nanomaterial-containing products.

The two statements in the previous section also shed light on assumptions behind administrative pragmatism's relationship between data and decision making. We see in the first statement Denison's apparent desire to close debate on nanotechnology risks. By arguing that such risks are urgent, he implies that the debate is over on whether such risks exist. The success of such an argument would neutralize the power of those who wish to keep the risk debate active. If the existence of risks from nanotechnology is "black boxed" as a scientific truth, then the field on which both scientific and political debate is played is significantly diminished, and the range of interests who are allowed to play in the debates is considerably reduced.[57]

Working under the same assumption that regulatory science data can bring risk debates to closure, Bergeson in the second statement keeps the nanotechnology black box open by arguing that risks must be tied to specific products. This implies that there ought not be a single nanotechnology black box, but instead thousands of nano-product black boxes, each needing to be filled with product-specific sets of data.

The conflict between these two perspectives on the use of data to close risk debates reflects administrative pragmatism's contribution to the scientism story line: maintaining discipline and order in a liberal democratic society through the use of publically sanctioned expert opinion and expert systems. While the conflict appears to be over impersonal institutional roles and data requirements, it in fact is a struggle over influencing human behavior. Porter's observation that measurement is really about disciplining people applies to risk information.[58] While both Denison and Bergeson see federal agencies and regulatory science data as central to addressing risk issues, they use language that suggests different strategies for influencing perceptions of nanotechnology risks: in the first instance, federal officials' perceptions; in the second, the public's. Denison's language suggests that federal agency behavior and public perception can be steered by emphasizing what data are presently available and what those data suggest for risks from nanotechnology generally. The bias in the language is toward action—that is, taking bold, broad action based on the existing data set. Bergeson's statement implies that more ordering, more information gathering, more *measurement,* to use Porter's language, is needed to inform narrower but more-numerous problem sets. While both individuals' statements embrace data-driven expert decisions, they clearly differ on how much data are needed (and therefore, how much uncertainty is tolerable) and how broadly those data should be applied to articulate positions on nanotechnology risks. As Scott has argued, those who define the quantification metrics define the risk debate, and discourse becomes framed around this definition.[59] For nanotechnology, the power struggle to define the nature of the relationship between data and risks has not been resolved.

What Metaphors or Other Rhetorical Devices Are Employed in the Statements?

The language of plans, strategies, systems, and the guiding role of science is prominent in administrative pragmatism. The following statement by Andrew Maynard concisely pulls together several threads of administrative pragmatism thinking.

> Safe nanotechnologies will not just happen. We will need leadership and guidance to help overcome our human scale perspective and ensure the rule book for safe nanotechnology is built on sound science.[60]

Maynard couples "leadership and guidance" with a "rule book" and "sound science"—all key words in the language of administrative pragmatism. Leadership is expected from two sources: government institutions and the institution of science. The rule book also has two meanings: rules promulgated by government, and following the rules of good scientific practice. Note also that these rules of safe nanotechnology are "built" upon a scientific foundation. It is not clear from the statement who the builders are, although within the context of administrative pragmatism, government would at a minimum be holder and enforcer of the rule book, which presumably would be written collaboratively between government and stakeholders. In the next statement, industry lobbyist Sean Murdock uses several words that reflect a perspective on how responsibilities would be shared among industry and government.

> Obviously businesses are responsible and accountable to make sure their products are safe. That is true for nanotechnology, that is true with everything. . . . The government, you know, we have talked about needs to develop the standards and the characterization protocols if you will to characterize these materials and the test methods to continue to evolve those based on the stated science to having our best understanding of what is in fact safe. And then industry needs to apply that.[61]

"Responsible and accountable" is applied here to industry, consistent with administrative pragmatism's emphasis on order and discipline. "Standards," "protocols," and "characterization"—all aspects of measurement and discipline—are central to calculated governance, and Murdock makes a clear delineation of who is responsible for what: government sets the standards, industry applies them to safe nanotechnology product development. An interesting deviation from the orderliness of administrative pragmatism is recognition that standards must "evolve" as the state of the science changes. This notion of evolution, and perhaps flexibility, is not prominent with the statements placed within this discursive category. Nevertheless, it may reflect the *pragmatism* aspect of administrative pragmatism: that is, the responsibility—and to some degree, willingness—to reconsider positions upon receipt of new information. In the statements I have located within administrative pragmatism, I find less rigidity or stridency than appears within risk society and technological progressivism language.

That said, the calculating nature of administrative pragmatism, empowered by scientism, exerts a discipline over risk governance that while perhaps not rigid is nonetheless powerful. To illustrate this point, I close this section with a quote from DuPont's Terry Medley as he describes the ED-DuPont Risk Framework:

> [The] objective [is] to develop and deliver a systematic and disciplined process for identifying, managing, and reducing potential environmental safety and health risks of engineered nanomaterials across all stages of a product's life cycle.[62]

This short statement captures both the essence of administrative pragmatism as well as the hold of scientism over the identification and management of nanotechnology risks. Within the nanotechnology risk discourse, scientism serves to discipline both language and practice.

SUMMARIZING DISCOURSE: WHAT NANOTECHNOLOGY RISK LANGUAGE TELLS US

One summary conclusion from this chapter is that ideas of order and power are prominent in nanotechnology risk language. The primary vehicle for creating or maintaining order and projecting power is regulatory science. Most striking is the universal acceptance of regulatory science's prominence by the individuals whose statements I have analyzed for this chapter.

Scientism is a story line that runs throughout nanotechnology risk discourse. One implication is the power of scientism to what I characterize as blur the boundaries between the risk society, technological progressivism, and administrative pragmatism categories. As has been discussed by numerous STS scholars and specifically here by Kleinman, scientism demands that science be the only legitimate means by which one may challenge a technology. Time and again we encounter statements in which scientism adds equivocation to language. This statement by Andrew Maynard shows scientism militating against a risk society message:

> Top of the list of bad habits is a tendency to treat risk-focused studies as economy-class research. Research into understanding and mitigating potential risks arising from emerging technologies is key to success in innovation. And the more innovative the technologies being developed, the more innovative the risk-research needed to use them wisely.[63]

Solely within the risk society discursive category, the main message of this statement would be that emerging technologies carry inherent risks, and that more-complex technologies create more-complex risk issues. And indeed the speaker says as much. However, the statement also argues that the way—apparently the only way—to address such risks is through regulatory science data. While I locate this statement within the risk society category because of the power of its risk inherency message, I also consider it to be an overlap statement with administrative pragmatism.

The next statement by Congressman Bart Gordon shows how scientism also adds ambiguity to technological progressivism.

> I feel there is no question that this committee understands the importance of nanotechnology, and recognizes the strong justification for a robust federal research investment. . . . Maybe there are no harmful effects. We simply do not know the necessary information to know if there are or aren't. What is clear is the commercialization of the technology is outpacing the development of science-based policies to assess and guard against adverse environmental health and safety consequences. The horse is already out of the barn. Thus, prudence suggests the need for urgency in having the science of health and environment implications catch up to or even better, surpass the pace of commercialization. We need to develop the tools and procedures to determine if nanomaterials are harmful, and if so, what specific controls may be needed.[64]

The Congressman's dominant message is, true to technological progressivism, the importance of advancing nanotechnology and ensuring that the technological "horse" is proceeding toward commercialization at full gallop. Yet "science-based policies" with the "tools and procedures" of regulatory science are needed to rein in any adverse impacts. This is another example of where it seems correct to locate this statement within technological progressivism, while noting how scientism causes it to overlap with administrative pragmatism.

A second important significance of the statements that overlap and blur the boundaries between discursive categories is their tendency to marry the power theme with other themes in the statements. In particular, the inherent power of nanotechnology, and power of the public to influence the success of nanotechnology are important themes within the three discursive categories.

Within the risk society statements, language that distinguishes the natural from the human is central to defining the discursive category. The statements concerning this relationship suggest that thinking about this distinction is unsettled—ideas oscillate between the nature-human divide and, consistent with the STS literature cited in this analysis, leave unclear whether the construction of such a divide serves or hinders the nanotechnology risk debate.

Technological progressivism centers around the relationship between nanotechnology's development path and societal progress. It is the most optimistic of the three categories of language, its optimism dampened only by the concern that a public backlash could lead to a derailment of nanotechnology as an engine of innovation and beneficial new applications and products. To keep nanotechnology on the right track, much technological progressivist language calls for the application of scientific information and communication strategies to educate the public on nanotechnology's benefits and, more broadly, to train them how to interact in an informed and supportive way to emergence of new technologies into society.

Accountability, responsibility, and the quest for data-driven decision making are prominent features of administrative pragmatism. This is the language of ordered, calculating democracy, where standards and methods are applied to regulatory science information to articulate nanotechnology risks in a manner that is transparent to the public and legitimate to stakeholders such as industry and environmental NGOs. Within administrative pragmatism, government is viewed as the primary authority for describing and managing risks; however, administrative pragmatism also makes clear that responsibility for developing safe nanotechnology is shared among all involved sectors of society.

Regulatory science is so dominant within nanotechnology risk discourse that it would appear from the statements analyzed for this case study that it is the only legitimate means for understanding risks. This has power-relations implications, in particular regarding the power of science and data to drive policy debates. These policy debates fit within larger contexts of calculating governance and global capitalism, which for nanotechnology are embraced as a positive forces by administrative pragmatism and technological progressivism; within risk society, the negative portrayal of these two forces is generally limited to concern that too much focus on the calculation of nanotechnology's potential economic benefits could lead to disputes similar to the GM foods debates in Europe. The following chapter takes a closer look at power relations, with particular focus on the interplay between calculating liberal democratic governance and global capitalism.

NOTES

1. Hajer, 49.
2. Dryzek, 9.
3. Hajer, 44.
4. Sara Mills, *Discourse* (New York: Routledge, 1997), 11. Like Dryzek, Mills discusses Foucault's use of discourse and its association with power, 16–22.
5. Hajer, 62–63. Also regarding story lines, see Saul Halfon, *The Cairo Consensus: Demographic Surveys, Women's Empowerment, and Regime Change in Population Policy* (Lanham, MD: Lexington Books, 2007), 21.
6. Jasanoff, *The Fifth Branch*, 5.
7. Dryzek, 15–19. Dryzek's sweeping analysis of environmental discourse nicely covers the broad spectrum of global perspectives on environmental protection. However, for my analysis of risk discourse in the United States, his analysis is too broad: for instance, it brings in the voices of survivalism and green radicalism which, while important to international environmental discourse, are not meaningfully present in US policy debates.
8. Risk society was the "quietest" discursive category because I located only eleven percent of all the statements analyzed for this study within this category. See Appendix, Figure 3A.
9. Nathan J. Comp, "Harmful or not? That's hard to tell: Nanotech watcher Andrew Maynard assesses its risks," *The Daily Page*, 5 July 2007, http://www.thedailypage.com/daily/article.php?article=7792.
10. US Congress, House of Representatives, House Committee on Science and Technology, Subcommittee on Research and Science Education, "Research On Environmental And Safety Impacts Of Nanotechnology: Current Status Of Planning And Implementation Under The National Nanotechnology Initiative," 110th Congress, First Session, 31 October 2007 (Washington, DC: US Government Printing Office), 123.
11. University of Montana, video: "Creating 'Good' Technologies? An Emerging Technology in the 21st Century" with Vicki Colvin Debating Science University of Montana Center for Ethics 2008 Summer Series. http://video.google.com/videoplay?docid=-8198115874611299878#.
12. University of Montana, video: "Creating 'Good' Technologies? An Emerging Technology in the 21st Century" with Vicki Colvin Debating Science Univer-

sity of Montana Center for Ethics 2008 Summer Series. http://video.google.com/videoplay?docid=-8198115874611299878#.

13. Donna J. Haraway, *Simians, Cyborgs, and Women: The Reinvention of Nature* (New York: Routledge, 1991), 9.

14. Donna J. Haraway, *Simians, Cyborgs, and Women: The Reinvention of Nature* (New York: Routledge, 1991), 3.

15. Donna J. Haraway, *Simians, Cyborgs, and Women: The Reinvention of Nature* (New York: Routledge, 1991), 209.

16. Bruno Latour, *We Have Never Been Modern,* translated by Catherine Porter (Cambridge: Harvard University Press, 1993), 141.

17. Bruno Latour, *We Have Never Been Modern,* translated by Catherine Porter (Cambridge: Harvard University Press, 1993), 97–99.

18. Bruno Latour, *We Have Never Been Modern,* translated by Catherine Porter (Cambridge: Harvard University Press, 1993), 97–99.

19. Bruno Latour, *We Have Never Been Modern,* translated by Catherine Porter (Cambridge: Harvard University Press, 1993), 104.

20. Bruno Latour, *We Have Never Been Modern,* translated by Catherine Porter (Cambridge: Harvard University Press, 1993), 91.

21. Bruno Latour, *We Have Never Been Modern,* translated by Catherine Porter (Cambridge: Harvard University Press, 1993), 32.

22. US Congress, House Science, 31 October 2007, 124.

23. US Congress, House of Representatives, Committee on Science, "Environmental and Safety Impacts of Nanotechnology: What Research Is Needed?" 109th Congress, First Session, 17 November 2005 (Washington, DC: US GPO), 68.

24. US Congress, House of Representatives, Committee on Science, "The Societal Implications of Nanotechnology," 108th Congress, First Session, 9 April 2003 (Washington, DC: US GPO), 49.

25. Comp, 5.

26. Andrew Maynard, "Ten things everyone should know about nanotechnology safety," Originally posted at 2020science.org.

27. Richard Denison, "A proposal to increase federal funding of nanotechnology risk research to at least $100 million annually," *Environmental Defense staff paper.* April 2005, 2.

28. Maynard, "Ten things everyone should know about nanotechnology safety."

29. Kleinman, *Science and Technology in Society,* 5.

30. Howard Lovy, Howard Lovy's NanoBot. "Being Mike Roco." 20 April 2004. http://nanobot.blogspot.com/2004/04/being-mike-roco.html.

31. Sarewitz, 19.

32. Sarah Graham, "Nanotech: It's Not Easy Being Green." Interview with Dr. Vicki Colvin. *Scientific American.* www.scientificamerican.com. 28 July 2003.

33. Miller and Rose, *Governing the Present,* 29–31.

34. US Congress, House Science, 31 October 2007, 87–88.

35. US Congress, House Science, 17 November 2005, 89.

36. Durkheim, 417.

37. Durkheim, 425.

38. Durkheim, 439.

39. Robert K. Merton, *The Sociology of Science: Theoretical and Empirical Investigations* (Chicago: The University of Chicago Press, 1973), 269.

40. Robert K. Merton, *The Sociology of Science: Theoretical and Empirical Investigations* (Chicago: The University of Chicago Press, 1973), 271. While there is an STS temptation to put scare quotes around *truth*, in this paragraph we are in Merton's world and therefore will leave the word as is.

41. Miller and Rose, *Governing the Present*, 15. Miller and Rose break Foucault's concept of governmentality into two aspects, rationalities/programs and technologies, describing the former as "a way of representing and knowing a phenomenon" and the latter as "a way of acting on it so as to transform it."

42. Shapin and Schaffer, 21.

43. Shapin, *The Scientific Life*, 5.

44. Shapin, *The Scientific Life*, 5.

45. Shapin, *The Scientific Life*, 311.

46. Durkheim, 417.

47. Lynn Bergeson, "Nanotechnologies and FIFRA." *ChemAdvisory*. vol. 35, July 2006: 4. This article was originally published in the April 2006 issue of the *Gradient Corporation EH&S Nano News*.

48. US Congress, House Science, 31 October 2007, 30–31.

49. US Congress, House Science, 31 October 2007, 18.

50. For example, see the *EPA Nanotechnology White Paper*.

51. Daniel Kleinman, *Politics on the Endless Frontier: Postwar Research Policy in the United States* (Durham: Duke University Press, 1995). The entire work covers this topic, but see in particular 92–99 regarding Bush's interactions with President Franklin Roosevelt. Also see Sarewitz, 102–104.

52. US Congress, House Science, 31 October 2007, 62.

53. Douglas and Wildavsky, 122–23.

54. Agamben, 82.

55. Lynn Bergeson, "Food and Drug Administration's Regulation of Nanotechnology," *Daily Environment Report* ISSN 1060-2976 BNA 9-22-08: B-1-B-6.

56. Ulrich Beck, *Ecological Enlightenment*, 53. This contention is disputed by advocates for environmental justice. See, for instance, Bernice Bovenkerk, "Is Smog Democratic? Environmental Justice in the Risk Society," *Melbourne Journal of Politics*, 2003. Full text available at http://www.thefreelibrary.com.

57. For a discussion of black boxing, see Bruno Latour and Steve Woolgar, *Laboratory Life: the Social Construction of Scientific Facts* (Los Angeles: Sage, 1979), 150.

58. Porter, 28.

59. Scott, 27.

60. US Congress, House of Representatives, Committee on Science and Technology, "National Nanotechnology Initiative Amendments Act of 2008," Hearing Before the Committee on Science and Technology, 110th Congress, Second Session, 16 April 2008 (Washington, DC: US GPO), 37.

61. US Congress, House of Representatives, Committee on Science and Technology, "National Nanotechnology Initiative Amendments Act of 2008," Hearing Before the Committee on Science and Technology, 110th Congress, Second Session, 16 April 2008 (Washington, DC: US GPO), 78.

62. Terry Medley, "Nano Risk Framework—An EHS Assessment Tool." PowerPoint Presentation, presented 20 March 2008, in Baltimore, MD, at a session titled, *Nanotechnology: The Future of EHS Regulatory Policy*.

63. Andrew Maynard, "Tough Love for Science and Technology Innovation." *Science2020*, blog posted on 10 December 2008, www.2020science.org.

64. US Congress, House Science Committee, 17 November 2005, 15–16.

4

Power and Its Grip on Environmental Discourse

Social forces exert power within discourse in ways that limit the range of voices heard in risk debates. In the United States, one such force is our calculating system of liberal democracy; another is global capitalism. In this chapter, I show how power is shaped by liberal democracy's system of calculated governance and global capitalism. My conclusion is that this shaping of power influences how risks are described, which in turn affects the quality of policy debates. For nanotechnology risk discourse to become more inclusive, these power implications must be understood, acknowledged, and addressed through specific policy actions.

Clearly there are interactions between risk, language, power, calculated governance, and global capitalism. As noted in chapter 1, the ideas discussed in this and the previous two chapters do not stand alone, but rather build on one another. The concepts of risk and regulatory science introduced in chapter 2 are further developed in chapter 3's discourse analysis, in particular with regard to the importance of regulatory science's role in carrying the scientism story line through nanotechnology risk discourse. In this chapter, the analysis of discourse in chapter 3 is further developed in terms of the impact of language and discursive practices on power relations. The chapter begins with a discussion of power, carrying over aspects of the previous chapter's discourse analysis to consider how power is manifest in nanotechnology risk language, in particular through the scientism story line. STS scholarship is brought into the analysis to help reveal political and power implications that are not apparent from the risk statements themselves. This discussion of power and scientism then is carried into an analysis of how our calculating democracy reinforces the privileged status of scientism and regulatory science in policy debates. The field of view is then widened to examine how global capitalism is an overarching force that locks calculation and science—*calculating science* that is used to produce quantitative estimates of risk—not only into liberal democratic society but more broadly into the vast system of human transactions and other interactions that create the processes and materials that generate environmental stressors. To conclude the chapter, this analysis is scoped back down to focus on what the nanotechnology debates tell us about how the power relations operating within calculated governance and global capitalism result in limiting environmental policy debates.

BIOPOLITICAL POWER

The power of government energy policy to influence commodity prices (for instance, by subsidizing corn ethanol production), or of advertising to create consumer demand for the newest electronic gadget, while relevant to the societal impact of technology, is not the kind of power that is the main focus of this study. In the nanotechnology environmental case, we are particularly interested in power's potential to control life and death. In modern liberal democratic societies, this power resides in the realm of politics and was labeled *biopolitics* by Michel Foucault.[1] Giorgio Agamben defines Foucault's concept of biopolitical power, or *biopower,* as "the growing inclusion of man's natural life in the mechanisms and calculations of power."[2] Foucault's work explores the relationship, within discourse, between knowledge and power in society.[3] The nanotechnology debates illustrate Foucault's observation that power and knowledge are imbricated such that the outcomes of power struggles determine what within discourse is considered knowledge.[4] As we have seen with chemicals and nanomaterials, regulatory science is the privileged provider of knowledge about impacts on the environment. Science, technology, politics, and human life since the Enlightenment have been inseparable and interrelated. However, they take on new, and greater, implications for the distribution and use of power in the twenty-first century; specifically in this study, with respect to the environmental impacts from emerging technologies. These new implications for biopolitical power are driven by the concept of risk.

BIOPOWER AND RISK

Risk, as the always-in-the-future possibility (or in regulatory science, *probability*) of harm, is directly related to power over life and death. In discussions regarding nanotechnology risks, biopower implications range from consumers' ability to know which products contain nanoparticles and therefore their ability to make choices on whether to use such products, to how the benefits and risks of nanotechnology might be distributed across society.

Biopolitics is being revealed in the nascent debates over possible environmental risks from exposure to nanomaterials.[5] New abilities to fabricate materials at the atomic and molecular levels add dimensions of possibility and complexity beyond those presented by traditional chemicals or materials. Developing within western society is what Jürgen Habermas calls a "de-traditionalization of a lifeworld," as nanoscale technologies operating in the realm of the unseeable (at least, without the use of sophisticated imaging technologies) present us with potential new, but largely unknown, concerns.[6]

Yet, despite all that is new about nanotechnology, power over nanotechnology risk discourse and decision making has been retained within what I would characterize as the *chemicals regime*—the apparatus employed by regulatory bodies in the United States and abroad, and supported by legislative and judicial frameworks and practices as well as by industry and NGO stakeholders, to govern the production and use of industrial chemicals. An indication of this control is the formation within the OECD's Chemicals Committee of the Working Party on Manufactured Nanomaterials, with a principal charge to the WPMN being to determine how the OECD's internationally harmonized chemical test guidelines can be used for environmental, health, and safety testing of nanomaterials.[7] In the United States, the EPA plays a leadership role in the administration of the chemicals regime, including nanomaterials. Within EPA, leadership on nanotechnology regulatory science issues has been shared by the Office of Research and Development, which funds through academic grants and conducts within its own laboratories research to support EPA's regulatory programs; and the Office of Chemical Safety and Pollution Prevention, which administers EPA's responsibilities for chemicals under TSCA and FIFRA.[8]

There are several biopower implications of the chemicals regime controlling risk issues. First, risk assessment for chemicals has been socially constructed to address human and ecological effects endpoints based on the societal repercussions of such events as the 1950s cancer scares and the 1962 publication of Rachel Carson's *Silent Spring*.[9] Specific health endpoints such as cancer and reproductive effects, and certain non-human organisms such as rodents and fish, are specified in guidance documents as the appropriate focus areas and biological systems for regulatory science data generation. Therefore, what environmental biopolitical impact means—i.e., the power of pollutants over life and death—is circumscribed by the socially constructed parameters of regulatory science data collection.

Second, the methods of risk science under the chemicals regime shape biopolitical power. The extensive and detailed risk assessment guidance documents cited in chapter 2 define and delimit what are acceptable knowledge-generation practices for identifying and characterizing risk, as risk is defined by regulatory science. The chemicals regime has an elaborate international scheme of test guidelines whose centerpiece is the OECD Mutual Acceptance of Data program mentioned in chapter 1. As a result, the developed nations that comprise the OECD in effect define the nature of chemical risk and determine how the international community creates knowledge about the risks of specific chemicals. In this way, the industrialized nations that produce the bulk of the world's chemicals retain power over determining the risks of those chemicals.

Finally, through its control over the definition of risk and how risks are measured, the chemicals regime wields power over who is considered to be a legitimate chemical risk expert. Jasanoff observes that regulators, the managers of the chemicals regime, view those individuals who are most conversant in regulatory science as being the most capable of judging the merits of chemical risk determinations.[10] Power therefore is further concentrated into a relatively small group of people deemed to be expert in chemical risks. Limitations placed by regulators on the expertise permitted to judge risk determinations complete the closed system of the chemicals regime: endpoints and species, data acquisition methods, and who is an expert define the boundaries of the acceptable knowledge within the chemicals regime. Since nanotechnology occurs within the chemicals regime, for nanomaterials the power over the definition and characterization of risks is as concentrated within regulatory science as it is for chemicals.

Yet the assertion in chapter 2 that regulatory science is a weak expert system raises the question of whether nanotechnology risk issues can remain confined within the chemicals regime. If nanotechnology truly represents a de-traditionalization of a lifeworld, then the potential may exist for a new examination of how society considers its environmental impacts. The context for such re-examination is already being built, as the US Congress considers a "Safe Chemicals Act" that would revise TSCA in ways that could open risk discourse to a broader range of voices. The statement by Senator Frank Lautenberg, the bill's sponsor, that "America's system for regulating industrial chemicals is broken," suggests that within the chemical regime's closed power system energy may be gathering for altering the governance of chemicals in ways that may shift existing power relations.[11] For instance, the bill states that it is United States policy to "promote the use of safer alternatives and other actions that reduce use of and exposure to hazardous chemical substances and reward innovation toward safer chemicals, processes, and products."[12] A new emphasis on green chemistry and safer alternatives implies a significant shift from current regulatory science emphases within the chemicals regime. Such movement toward Chapman's "riskiness" perspective could change the biopower dynamic by requiring more questioning of a chemical or nanomaterial's potential for impact at the design or development stages, rather than waiting for the potential for an impact to occur before performing an assessment of risk. The economic stakes for commercializing a chemical are lower at these earlier stages, which may lead industry and regulators to be more open to broader considerations of what risks mean and how they should be addressed.

Chapter 3's discourse analysis also reveals an aspect of the risk-power relationship that may challenge the capability of the chemicals regime to apply existing regulatory science practices to nanotechnology: concern that a public backlash could preempt regulatory science and lead to a societal rejection of nanotechnology based on the public's own view of risk. Several statements caution against the possibility of another GM food episode, referring to the European public's rejection of genetically modified food crops.[13] However, in discussing agricultural biotechnology, Kleinman argues that technology development tends to follow the trajectory historically followed by a given society.[14] Prior to the introduction of transgenic crops, American farmers had already followed a development path that made them much more reliant on technological inputs than were their European counterparts. Indeed, American society in general had followed a development trajectory more-closely aligned with new technologies, and therefore the American farmers and public were more receptive to genetically modified foods than were Europeans.

Just as transgenic crops were for Americans an extension of an already technology-dependent agricultural sector, so too are nanomaterials likely to be viewed as an extension of chemistry. Given the current dominance of regulatory science in the nanotechnology risk discourse, and the origin and power center of that discourse being the chemicals regime, it is unlikely that society will disassociate nanomaterials from chemicals. More probable is that, like chemicals, the public will maintain an ambivalent relationship with nanotechnology, appreciating its ability to enhance the products that society desires but worrying about downstream health and environmental impacts. If this is so, then the risk-power relationship for nanotechnology will evolve as broader public debate on chemicals evolves. The outcome of TSCA reform efforts, exemplified by the Lautenberg bill, may signal whether the chemicals regime is willing, ready, and able to consider risk differently than it does presently. Whether or not the regime evolves to accommodate broader considerations of risk, the public's continued association of nanotechnology with chemicals makes a public backlash against nanotechnology seem unlikely, given American society's dependence on chemicals for its perceived economic and social well-being.

NANOTECHNOLOGY RISK DISCOURSE: BIOPOWER, RISK, AND CONTROL

Those who control the language control the power. This seems an aphorism as we witness events in our daily lives and on the world stage. In our workplaces—in offices and meeting rooms, or on the shop floor or out on the construction site—professional conversation is circumscribed by the objectives of the organizations that employ us. Two farmers discuss the price of corn, and it is a conversation framed largely by global commodity markets. When a group of government bureaucrats meets to consider regulatory options, their discussion is bounded by the statutory framework within which their agency draws its administrative authority. The ever-shrinking number of us who watch the televised evening news see world leaders frame issues in terms of economic prosperity and national security, as if those two concepts covered the totality of human existence. Advertising of all types makes the case for the benefits of one product over another, with no consideration as to whether such a product is needed at all. Language thus draws boundaries around the limits of socially acceptable action, and the establishment of limits is an exercise of power. Karen Litfin argues that discourse puts boundaries around not only actions, but also around the limits of thought itself.[15] This implies a type of power that extends beyond individual agency, to the collective social force that joins together individuals within discourse.

The dominance of the scientism story line and regulatory science within nanotechnology discourse indeed puts boundaries around thinking about risk, and these limits are discussed in the previous section. However, even broader than the risk-power dynamic are the power relations that result from humans interacting with the rest of the world. Chapter 3 highlights the importance of the nature-culture, or the natural-"man-made" divide, in nanotechnology risk statements. These natural-human relations are important to the nanotechnology debates for three reasons relevant to this chapter's focus on power.

First, risk in the environmental protection context typically refers to impacts of anthropogenic origin. While living things certainly are exposed to dangers of non-human origin, such dangers typically are accepted as a consequence of life and are not subject to environmental regulation or risk mitigation.[16] A manufactured carbon nanotube, for example, is given closer scrutiny than a similar-sized particle of dust. Because of the consequences associated with attaching risk to an object or action, there is power in the ability to articulate through language what is natural and what is a technological artifact. The power implications relate, at one level, to responsibility for identifying and defining risks. The following statement, framed in the language of regulatory science by an industry coalition called the Silver

Nanotechnology Working Group, exemplifies an attempt to gain definitional control by exploiting ambiguity between whether nanomaterials are different from naturally occurring compounds:

> The one hundred-plus year history of human exposure through medicinal nanosilver colloids, the fifty-plus year history of FIFRA-registered swimming pool and spa algaecides, the forty-plus year history of FIFRA-registered nanosilver impregnated water filtration, the ten-plus year history of nanosilver impregnated FDA-approved medical devices—all with no reports of significant incidents—is ample proof that no unreasonable exposures and/or risks are posed by nanosilver.[17]

In other words, silver has been with us for a long time, and silver nanomaterials present no new issues for society.

Second, our calculating democracy seeks to quantify and bring order not only to how people interact with one another, but also to human activity in relation to everything else. For instance, measurements such as temperature, noise level, and air quality are taken and characterized in terms of their human relevance (e.g., hot, loud, smoggy) and then projected through language and practice with the implication that their meaning has relevance broadly in the environment. The word *nanometer* is given significance not only in terms of its definition as one-billionth of a meter, but also by the properties observed by humans, and deemed relevant by humans, of materials created in the nanoscale range. The power to link measurements with meaning in discourse is the power to determine what is in and what is out of consideration for environmental relevance.

Finally, global capitalism appropriates nature to meet human needs, and prices nature in accord with the value associated with nature's contribution to meeting human needs. Pricing is a kind of language, communicating cost and value. In nanotechnology, quantum effects exhibited at the atomic level—the finest level of division, measurement, and quantification to which humans have been able to subject nature—are harnessed by capitalism to create new wealth and strengthen capitalism's hegemony over human activity. The ability to articulate the cost and value of human control over material properties through nanotechnology thus becomes the ultimate reductionist strategy in legitimizing the extraction of resources from nature for human gain.

The STS literature complements discourse analysis by offering important observations on the natural-human relationship with respect to biopolitical power, and so it is useful to return to some of the STS literature introduced in chapter 3. While both Donna Haraway and Bruno Latour see the nature-culture divide as socially constructed, an important contrast between their perspectives is how they characterize the intent behind the use of power. Haraway's constructivist theory positions power as a tool for maintaining cultural domination. In Latour's account, power exerts a *shaping* force on the construct of the divide, but more in terms of mobilizing actors than in maintaining social positions.

The idea of acquisition is central to Haraway's argument. Nature is *appropriated* by western society for the purpose of *naturalizing* the mythology of human development.[18] That is, if science can demonstrate that the West's dominant cultural attributes are "natural," then the argument for maintaining status quo power relations is reinforced. For Haraway, the use of science for acquisition purposes is consistent with science's coextensive development within western capitalism. Science, naturally, would be structured in a manner consistent with capitalism's emphasis on management, control, competition, and resource acquisition and exploitation.[19] Within the nanotechnology risk discourse, we observe repeated use of language about "control" at the nanoscale.

Latour sees paradox as the principal implication of the constructed nature-society divide. Modernists set nature apart by identifying nonhuman Laws of Nature, yet they create these laws within their laboratories. The experiments are clearly constructed, yet the modernists maintain that their experimental results are not. Concerning chapter 3's discussion of hybrids, Latour asks how nature can be considered unconstructed, yet be constructed in the laboratory.[20] The implications of Latour's hybrids relate to the problems of power and domination that concern Haraway. For instance, Latour's model suggests that the addition of feminine voices makes us even less modern. While the modern perspective would purify the separation between nature and humanity/society, feminist history—by acknowledging and validating the feminine voice in science—adds yet another set of hybrid possibilities for the blending of nature and human society. Incorporating feminist history into Latour's notion of hybrids, one could argue that we humans were never modern, in part, because in modernity women have always been considered part of nature and yet also always have joined men in the pursuit of gaining knowledge. The work of Carolyn Merchant illustrates this paradox for modernity.

Carolyn Merchant uses historiography to evaluate the impact of Francis Bacon and others on western society's domination and exploitation of nature. What Merchant brings to this discussion of power is a recognition that society's relation with scientific knowledge has specific historical roots that run through deep-seeded gender and cultural terrain. The Baconians used science to advance the drive for power over nature, and use of the female image was one approach to applying the scientific method to human domination of the natural world.[21] Sexual imagery used by Bacon promoted the forceful extraction of nature's secrets and later, Robert Boyle's work incorporated into the experimental method the notion of nature as a woman who could be "controlled and dissected" in the laboratory.[22] For

Merchant, though, what ultimately killed nature was not gender bias, but rather the *mechanism* embraced by the new science: the idea that the world is a machine that can be broken down into increasingly smaller parts, ultimately to the atomic level, and that knowledge can be "abstracted from nature" by studying those parts and their relationships to the whole.[23] With nature reduced to a soulless machine, man can with a clear conscience fully exploit the world. Merchant sees the hope for nature's resuscitation in the development of an ecological view of knowledge that, because of its holistic perspective, begins to dismantle science's mechanistic superstructure.[24]

The discussion of the Haraway, Latour, and Merchant views on the power relations associated with the natural-human divide (if such a divide exists) has clear relevance to biopolitical considerations within the nanotechnology risk discourse. The language of control and appropriation, and the doubt that hybrids and feminist voices cast on the legitimacy of science's role in supporting human dominion over nature, all apply to a discourse on nanotechnology risks dominated by regulatory science but also infiltrated by language of uncertainty about whether science can adequately address concerns over nanotechnology specifically and emerging technologies in general. In the following sections, calculated governance and global capitalism are examined as social forces that exert power and control within the nanotechnology risk discourse.

CALCULATED GOVERNANCE AND THE CALCULATION OF RISK

Science underpins quantification and calculation. The success of calculated democracy in bringing political order to a decentralized, individualistic America is nothing short of remarkable. We rally around the Dow Jones Industrial Average as if it were our national home team, following its ups and downs like loyal Mets fans. We treat the unemployment rate like socioeconomic body temperature, with somewhere between four to five percent being our national 98.6°F. We have been made into African Americans, Asians, Caucasians, and Hispanics—each with our own statistics, which are uniquely American statistics. For better or worse, science lends authoritative support to these measurements, and these measurements indeed hold great authority. American policy makers and their relatively weak bureaucracies must demonstrate, through numbers, the rationales for their policy actions. These numbers in turn engender governmentality technologies perhaps not envisioned by Foucault, such as public interest groups and their boundary-probing strategies and techniques. Porter notes that our numbered society has proliferated experts outside government, and that these experts can serve as rivals to counterbalance state power through their challenges to government calculations.[25] Hence in America, think tanks and NGOs abound and are part of the public policy arena.

For liberal democratic governments, political questions are at their root about the ordering, controlling, and sustaining of human life. This last characteristic, sustaining life, makes them biopolitical questions. Increasingly since the middle of the twentieth century, science and technology have impinged on the politics of life. As stated by Nikolas Rose, modern science and technology have allowed us "to control, manage, engineer, reshape, and modulate the very vital capacities of human beings as living creatures."[26] There are characteristics of calculated governance that are shared among liberal democratic societies. Prominent among these is the relationship between calculation and power. Ezrahi makes a strong case that in western democracies, science supports the state in maintaining order, both real and perceived, while at the same time helping the state support its claims of being transparent and accountable.[27] An example is the US Bureau of Labor Statistics' employment figures. Its March 2009 *Employment Situation Summary* provides detailed tables with numbers (e.g., Hispanic labor force at 22,188,000) that give the sense of precise government knowledge of the nation-wide workforce dynamic.[28] This detailed, transparent statistical information provides both accountability and an orderly accounting of an important aspect of the economy. Government, as generator and communicator of employment statistics that individual citizens believe and internalize into their personal evaluations of the economy, holds considerable power over the collective representation of this important aspect of society. In this manner, science is instrumental to maintaining a sense of state control.

For nanotechnology, calculated governance is central to the development and application through regulatory science of the language of quantitative risk assessment in communicating perspectives on the potential environmental risks of manufactured nanomaterials. Numerical values serve both to order nanomaterials within specific risk contexts as well as, using dollar values, to describe their benefits to society. However, the language of calculation related to nanotechnology is at this point largely non-quantitative, reflecting uncertainty both in how much economic benefit is being and can be derived from nanotechnology, as well as to what the scientific data indicate regarding risks as defined by regulatory science. The following statement by Vicki Colvin illustrates the power of calculation:

> Whether to slow down the pace of a new technology is a divisive question for society. In the case of nanotechnology, the question of government regulation will be contentious, as known benefits must be balanced against an incomplete view of the risks.[29]

Here technologist Colvin stacks the calculation deck in favor of nanotechnology without needing to use numbers. She identifies the benefits of nanotechnology as "known" and capable of being "balanced" (as on a scale) with risks of which we have an "incomplete view" presumably because they are not articulated and quantified in the language of regulatory science.

Richard Denison is no technological progressivist and the following statement from his 2007 hearing testimony pushes back against Colvin's notion of the value of calculation. At the same time, however, Denison's statement is very much a part of nanotechnology's regulatory science narrative.

> At this stage, it is not possible to apply a typical "value of information" methodology to predict what type of research will optimally reduce uncertainty about risks or lead to broader knowledge and understanding of nanomaterial behavior without broadly speculating on potential risks and the types of information needed to reduce them. Value of information methodologies rely on quantifying the harms being reduced, which is not possible at this time for nanomaterial risks. Moreover, in the setting of an emerging technology such as nanotechnology, the economic consequences of obtaining or failing to obtain critical information on toxicity are in reality so unpredictable that formalizing the costs and benefits through a value of information analysis is artificial and potentially misleading.[30]

Denison sees nanotechnology's benefits and risks as being equally incalculable at present, with any attempts to do such calculations possibly leading to "misleading" conclusions. However, although he is reluctant to support benefit or risk calculation, Denison's reticence is due solely to what he perceives as a lack of data to support such calculations. His statement suggests that he is as strong a supporter of regulatory science developing data to support calculation as is Colvin—perhaps more so, since while Colvin even with little data is prepared to reach at least some conclusions about which direction the nanotechnology benefit-risk balance tilts, Denison withholds judgment until the data have been developed.

Denison's greater need than Colvin's for data-driven calculation is not surprising. Colvin, as a technologist and academic laboratory scientist, is less closely aligned with the policy making apparatus than is Denison as a member of an environmental NGO. Like private industry, and in general unlike academic scientists, NGOs engage with policy makers in making and describing calculations about environmental risks. But there is a distinction. NGOs move much more quickly than does private industry to converting what Ezrahi calls *high-cost realities* into *low-cost realities*. High-cost realities are data-rich descriptions of the world (in the nanotechnology case, descriptions of environmental risks and/or economic benefits), elaborately constructed through great effort and cost; low-cost realities are simplified representations of the world that are accessible to the general public.[31] In environmental policy, NGOs translate high-cost calculations into low-cost risk statements amenable to public consumption through blogs, newsletters, press statements, and news articles. Private industry, with greater resources to apply to direct engagement with regulators, typically will interact with government officials in the realm of high-cost reality, only translating risk information into low-cost forms if irreconcilable differences with regulators or public relations exigencies force them to issue public communications. Regulators and industry officials rarely if ever blog about risk; NGOs do so frequently. In general, construction of high-cost environmental risk calculations could be considered less democratic and transparent than is low-cost construction. However, risk calculations translated into low-cost statements of reality tend to lose some of the rigor and objectivity that Porter argues is crucial to government's legitimate biopolitical discipline over public health and well-being.[32] In this regard, Ezrahi issues a warning with his discussion of *outformations*—outward flows of representational images of the world. Because outformations are not created with any transparent methodologies, they are unavailable for public analysis. He suggests that low-cost outformations indicate a postmodern shift from reasoned discourse toward speculative decision making.[33]

Nevertheless, so entrenched is the need within liberal democracy for objectivity claims to be made through calculation that NGOs, industry, and government appear to play by the same linguistic rules when discussing the environmental risks from nanotechnology. What follows are three statements about one type of nanoparticle, titanium dioxide (TiO_2), from each of these groups; their treatment of risk calculation is quite similar, even if they reach somewhat different conclusions. The first statement is by the ETC Group, an NGO which in 2003 called for a moratorium on nanotechnology research and development.

> The crux of the issue is that a reduction in size—with no change in substance—can make a substance stronger or more reactive or lighter or more water-soluble or more heat-resistant or a better conductor of electricity. Property changes begin to happen with materials 100 nanometers or smaller (a nanometer is one billionth of a meter). It is these "quantum effects" that make nano-scale materials interesting to scientists and potentially profitable to industry, who are taking advantage of unique property changes in order to create new products and new markets. It should be no surprise that toxicity is another property that can change with a reduction in size: a chemical compound at the micro-scale—titanium dioxide (TiO_2), for example—may be benign, but a nanoparticle of that same TiO_2 could be toxic.[34]

Here the ETC Group follows what has become the common approach of acknowledging nanotechnology's potential benefits before launching into a discussion of risk. Even a group that has been consistently critical of what it perceives as a non-precautionary approach to nanotechnology feels it necessary to acknowledge possible economic benefits. Shifting from benefits to risk, the statement's assertion that "it should be no surprise" that a compound's risk profile may change through its fabrication at the nano scale is an example of a low-cost construction of risk. What makes this statement an outformation is that it glosses over the great uncertainty that exists over the relationship between the size of a particle and its potential for environmental impact. It appears that through this statement the ETC Group is attempting to black box the relationship between particle size and toxicity.

The next statement is excerpted from a literature review of TiO_2 nanoparticle toxicity studies within the final report of the Engineered Nanoparticles–Review of Health & Environmental Safety (ENRHES) project. The ENRHES activity is interesting because, under the European Commission's Seventh Framework Programme, it was carried out through a collaboration of academic, government, and industry scientists.

> The results of the studies outlined imply that the toxic potential of TiO_2 was primarily dictated by particle size, and crystallinity; whereby decreasing particle size, and anatase forms of TiO_2 enhanced particle toxicity. In addition, it is suggested that the exposure scenario (including species used, exposure method and dose administered) was able to impact on the toxicity of metal oxides, which complicates making comparisons between investigations.[35]

The ENRHES statement differs from that of the ETC Group because it communicates the high-cost information (an extensive literature review) contained in the project report. While like the ETC Group statement the ENRHES report associates particle size with toxicity, it also adds to the risk calculus crystalline structure and exposure considerations. A US federal government report by EPA takes high-cost calculations even further in its characterization of TiO_2 nanoparticles.

> Not all nano-TiO_2 is the same. Commercially available brands of nano-TiO_2 can vary in particle size, surface area, purity (e.g., due to doping, coating, or quality control), surface characteristics, crystalline form, chemical reactivity, and other properties (see Table 1-1). Nano-TiO_2 is available in pure anatase, pure rutile, and mixtures of anatase and rutile. In general, anatase nano-TiO_2 is more photocatalytic than the rutile form, and nanoscale rutile is less photoreactive than either anatase and rutile mixtures or anatase alone (Sayes et al., 2006). However, a mixture of 79 percent anatase and 21 percent rutile nano-TiO_2 (P25) was found to be more photocatalytic than 100 percent anatase nano-TiO_2 in some instances (Coleman et al., 2005; Uchino et al., 2002), but less effective in others (Nagaveni et al., 2004). Such contrasts point to the role of other factors in accounting for the behavior and effects of nano-TiO_2. For example, surface treatment of nano- TiO_2 can change nano-TiO_2 activity, including photoreactivity. Aeroxide T805, which is nano-TiO_2 that has been treated with trialkoxygoctyl silane on the surface, has very low surface reactivity

> (Degussa, 2003). Similarly, surface coatings of silicone and other compounds are used to decrease nano-TiO_2 photoreactivity so that nano-TiO_2 can be used to protect human skin, plastic, and other objects from UV radiation.[36]

These three statements illustrate increasingly high-cost constructions of risk calculation as one moves from NGO, to academic/industry, to government regulatory statements of nanotechnology risk. It is not surprising that documents produced by regulatory bodies would go to the greatest lengths in communicating complex calculation. Ezrahi points out that constructing political order involves using science to depersonalize the use of political power. Such depersonalization is crucial to addressing democracy's inherent conflict between individual freedom and public welfare. Through the mechanical objectivity of science—in the nanotechnology case, regulatory science—policy makers are alienated from their actions. That is, personal motive is detached from their actions.[37] By objectifying policy actions in quantified language that has been co-produced in society, this alienation through depersonalization reinforces the concept of representative government, a key governance infrastructure that stabilizes the countervailing forces of individualism and public good. The releasing of information considered objective into society for analysis and critique also strengthens the coextensivity of private will and public good, by providing a means for individual evaluations of, and challenges to, government actions.[38]

Each of the above statements, in its barest sense, is addressing the same biopolitical question: What determines whether the introduction of TiO_2 nanoparticles into society is likely to harm people or other living things? The "what" consideration is inherently both calculating and political. It is inherently calculating because all participants in the nanotechnology risk discourse accept regulatory science as the means for articulating risk, and regulatory science is by its nature calculating. It is inherently political because participants in the discourse engage in power struggles over how calculations of risk are communicated through democratic society. At present, because of regulatory science's firm grip over discourse on nanotechnology's environmental implications, the balance of power in the struggle for control over nanotechnology calculation tilts to those constructing high-cost constructions of the risk reality. Those with the power to make compelling—that is, elaborate and robustly quantitative—calculations of risk hold the biopolitical advantage in shaping nanotechnology risk discourse.

GLOBAL CAPITALISM

Encircling America's framework of calculated governance is the Weberian hard shell of global capitalism. Capitalism as a vehicle for trans-global industrialization is a prominent feature of the nanotechnology risk discourse. Global capitalism surrounds not only the calculated governance of liberal democracies, but also societies that are neither liberal nor democratic. China and Russia, for example, are directing significant resources toward developing nanotechnology-related industries.[39] As Michael Hardt and Antonio Negri note, when imperial sovereignty expands, it does not destroy the cultures or powers it faces but instead subsumes them into what the authors see as a network of imperial sovereignty, but what I view as an intellectual framework and operational network of global capitalism.[40] Capitalism, then, continuously evolves as it absorbs attributes from the powers that it overcomes and incorporates into its global network. Therefore, it is not sufficient to view the influence of capitalism on the nanotechnology risk discourse solely from the perspective of liberal democracy. Rather, we recognize that environmental policy discourse generally, and nanotechnology risk debates specifically, increasingly are acted on by forces within capitalism that either are exogenous to liberal democracy (such as Chinese industrial policy) or straddle liberal democratic and other societies (such as advertising and trade agreements). The creation of demand for new technologies and products, and the building of industrial capacity to meet that demand, has become a unifying logic guiding the actions of institutions and societies around the globe. It therefore cannot avoid influencing scientific inquiry, including regulatory science, and thereby impact the nanotechnology risk discourse.

Herbert Marcuse, within his argument that advanced industrialization has flattened societal thought to a singular conception of rationality, views science's attempts (at the bidding of industrialism) to conquer nature as being motivated by a larger totalitarian goal of subjugating humanity.[41] Marcuse wrote his work during a time of struggle between capitalist and communist economic ideals, but for him practitioners of both ideologies operated under the larger totalitarian force of modern industrialization as the single organizing rationality for human behavior.[42] Now that communism for the time being appears to have lost that struggle, the argument for the role of science in advancing the rationality of industrialization can be centered around global capitalism. Marcuse's writing pre-dated Foucault's articulation of biopolitical power, but with respect to the nanotechnology case we are talking about the same thing: how global capitalism as a social force wields power over environmental risk debates. Discourse analysis is useful for identifying such power relations.

Global capitalism is prominent within nanotechnology risk discourse. In language within the discourse, capitalism typically is characterized in terms of US economic position; in particular, how nanotechnology might help grow the US economy, advance the nation's competitiveness in international trade, and enhance American leadership in the development, production, and marketing of new technologies. Statements such as the following by the Society for Manufacturing Engineers reflect this aspect of the nanotechnology risk discourse:

> Molecular nanotechnology promises to usher in the next Industrial Revolution and replace our entire manufacturing base with a new, radically precise, less expensive, and more flexible way of making products. These pervasive changes in manufacturing will leave virtually no product, process, or industry untouched. To be sure, nanotechnology has the potential to disrupt entire industries while leading to the creative destruction of current business models. It has already dramatically changed the competitive landscape of many industries worldwide including advanced materials/composites, aerospace/defense, automotive, energy, life sciences, medicine, electronics and semiconductors.[43]

In the context of "the next Industrial Revolution" and the rationality of global capitalism, risk issues often are framed negatively—discussing risks gets in the way of nanotechnology's potential to advance capitalist objectives. However, there also are examples of statements that portray risk issues as advancing global capitalism. For instance, using risk analysis to give a nanomaterial a clean bill of environmental health is seen as an effective strategy for gaining consumer acceptance of nanotechnology.[44] In other statements, competitive commercial advantage is thought to accrue to those countries that can use nanotechnology in products that are less-risky substitutes for traditional chemicals and materials. Capitalism thus is a strong motivator for participation in the nanotechnology risk discourse. While public-good applications of nanotechnology, such as its possible use in medical treatments and environmental clean-up, are mentioned in language articulating the nanotechnology benefit-risk calculus, discussions of these uses are side shows to the center-stage polemic on the introduction of nanomaterials into industrial and consumer products. In modern liberal democracies, or for those countries that wish to trade with modern liberal democracies, regulatory approvals are hurdles to be jumped in the race to product development and commercialization.

Therefore, much of global capitalism's power over nanotechnology risk debates is exercised in areas related to data generation for regulatory science. Who provides and manipulates the data used in regulatory science has important biopolitical implications; likewise, power structures, including legal frameworks and the allocation of resources across society, influence who is able to generate and use such data. For instance, the competitive nature of global capitalism leads to the need to maintain trade secrets, which results in much product data submitted to regulatory agencies being claimed as confidential business information and therefore inaccessible to the general public. Elements of the system of global capitalism—including multinational business operations; the transnational movement of human resources, in particular scientists and technologists; and international environmental treaties, harmonized chemical test guidelines, and technical standards—all impact the development and use of regulatory science, as well as how power structures in the United States react to and accommodate positioning around the world regarding the development and use of nanotechnology.

This history of nanotechnology research and development funding by national governments illustrates the strong link between global capitalism and public policy. Since 2000, the United States has tripled, and Europe and Asia have quadrupled, their expenditures in nanotechnology R&D. Total R&D expenditures from these countries/regions, which include government, corporate, and venture capital funding, have risen even more rapidly, in particular in Asia where they increased over six fold from 2000 to 2008.[45] The President's Council of Advisors on Science and Technology speaks in its 2010 report of the United States having "lost ground to foreign competitors" regarding nanotechnology funding and other aspects of the technology's development.[46] What began as laboratory-based inquiry into the ability to manipulate matter at the nanoscale has developed into an international competition to leverage public research to develop new national sources of economic growth.

In my analysis of the nanotechnology risk discourse, over 100 of the 665 statements in some way refer to global capitalism. Capitalism is explicitly prominent in statements falling under the technological progressivism discursive category, primarily through statements by industry and government representatives. However, capitalism is not absent from language falling under the risk society or administrative pragmatism categories. Within the risk society category, the primary concern is that capitalism has de-centered people from discussions about risk, their having been elbowed out by geopolitical factors centered around economic competitiveness. The following statement by Andrew Maynard nicely captures this sentiment.

> Research programs, strategies and policies are increasingly being influenced by people who lack a professional cultural bias toward focusing on the individuals their work and decisions will affect. That is not to imply that these people do not care—in many cases, they clearly do. But without that ingrained culture of putting others first, I wonder whether there is a danger of nanotechnology risk research being driven more by political expediency and the promise of economic gain, and less by the need to protect people. If this isn't the case, I am willing to stand corrected. But if it is, we need to work out how to get *people* back at the center of the nano-risk enterprise.[47]

In the risk society, global capitalism generally is portrayed as a potentially negative force. The perspective is nearly the opposite within technological progressivism and administrative pragmatism, where the identification of nanotechnology risks is viewed as antagonistic to a positive relationship between technology development and global capitalism. Within both of these discursive categories, statements that address global capitalism ascribe power to the public to make or break nanotechnology as an engine of economic growth. Speakers tend to hark back to European rejection of genetically modified foods as a situation to avoid with nanotechnology. However, there are subtle differences between how within technological progressivism and administrative pragmatism, speakers articulate the public's power over technology acceptance, and looking at a statement from each category can illustrate the differences.

The following statement from Sean Murdock of the NanoBusiness Alliance testifying before Congress represents a technological progressivist view of the relationship between risk and global capitalism:

> The general public still does not have a solid understanding of nanotechnology, despite the best efforts of the Alliance, its members, and countless educational institutions throughout the country. The bill takes some steps to help address this situation, which is important because the public's lack of a clear understanding of nanotechnology is one of the greatest risks that the nanobusiness community faces. International cooperation is critically important, especially in the area of standards development—but when it comes to educating the public, I believe that we need to be educating Europe rather than asking for their help to educate us. It seems that not a month goes by without an over-hyped scare story from Europe, or another argument for the precautionary principle in the EU. Coverage of and education about nanotechnology in the United States is much more balanced and takes into account the very real benefits of nanotechnology even as it speculates about risks.[48]

Here we see a deficit-model approach to the public's perceptions of risk: if only they had more information, the public would come around to supporting nanotechnology as an engine of economic growth and societal benefit. The following statement from Vicki Colvin provides a perspective more-reflective of administrative pragmatism's view of scientific data as a foundation of sound decision making.

> The public backlash against genetically modified organisms (GMOs), which detractors labeled "Frankenfoods," crippled the industry and ultimately cost billions in lost future revenues. The campaign against GMOs was successful despite the lack of sound scientific data demonstrating a threat to society. The founders of the Human Genome Project did not try to bury these legitimate concerns by limiting public discourse to the benefits of this new knowledge. . . . And in the case of nanotechnology, the ultimate losers may be the American taxpayers who invested over one billion dollars in nanomaterials research without any hard data on their toxicological and environmental effects.[49]

For academic and technologist Colvin, regulatory science data are a hedge against the main investment strategy of advancing technology to fuel economic growth. The "hard data" on environmental impact can play hedging roles in service to industrialization and capitalism in at least three important ways. First, risk data can be used by industry and their insurers to price potential liabilities (for instance, from workers inhaling asbestos-like carbon nanotubes) into product-development strategies. Such pricing allows industry and their insurers to spread risk over time (by setting aside capital for future claims) and over space through the reinsurance market.[50]

Second, data on material properties associated with risk can be factored by companies into green chemistry decision making. For instance, if regulatory science data suggest that exposure to fiber-like carbon nanotubes may lead to the types of lung disease seen from asbestos exposure, then companies may be able to produce modified carbon nanotubes that still provide the needed product performance but do not induce lung disease effects in laboratory animals. This provides an opportunity to "black box" carbon nanotube risks and produce a product deemed at least not like asbestos, if not entirely safe. If the product containing the carbon nanotubes has strong enough consumer pull (say, for instance, a better-performing golf club), being "safe enough" may be adequate, particularly if potential risk appears to be limited to workers and in particular if those workers make their contribution to global capitalism from countries that are not highly regulated and politically transparent liberal democracies.

Finally, the inherent uncertainty associated with regulatory science information, particularly when compared to relatively clearer prospects of benefits, provides a hedge for global development and commercialization of nanotechnology. Daniel Sarewitz notes that in our capitalist society, we accept non-scientific reasons for adopting technologies and their concomitant risks, but we require scientific—that is, risk-based—justifications for preventing or limiting the use of technologies.[51] Intentionally or not, we have set up our society to readily embrace new technologies and have constructed both science and science's relation to our liberal democratic institutions in such a manner that the risk game is rigged in favor of advancing industrial capitalism. Part of the reason why the risk game is rigged, as discussed in chapter 2, is that regulatory science is a weak expert system fraught with uncertainty, particularly regarding estimates of risk from nanomaterials. The forces aligned with global industrialization and capitalism are well positioned to exploit this uncertainty, since science is designed to falsify, and not to prove, hypotheses of risk.[52]

Regulatory science purports to be about imbuing decision making with data-driven rationality, yet the problem of induction presents a very practical concern for those attempting to support their policy decisions with scientific information. The issue is that the accumulation of scientific data rarely, if ever, supports theories of causality between exposure to a particular contaminant and adverse health or ecological outcomes to a degree sufficient to satisfy those who have a stake in the causality debate. Underdetermination, in the form of uncertainty in risk assessment, is used as a hammer to chip away at an opponent's position. While overt attacking of risk positions has not yet occurred with nanotechnology, since there are not yet sufficient data from which to construct robust assessments, such attacks have occurred regularly within the chemicals regime. This is illustrated in the following account from a 1995 congressional hearing on risks from pesticides. This particular exchange centers around whether to repeal the Delaney clause, which held that no food additive would "be deemed safe if it is found to induce cancer when ingested by man or animal" and therefore a safe maximum residue limit, or tolerance, could not be set for that chemical additive.[53]

In this hearing before the House Commerce Committee's Subcommittee on Health and the Environment, Representative Michael Bilirakis from Florida in opening remarks pointed out that since enactment of the Delaney clause, scientists had been able to detect increasingly smaller amounts of residues, down to parts per trillion. Such small amounts, argued the congressman, "pose no hazard whatsoever."[54]

Jay Feldman from the National Coalition Against the Misuse of Pesticides responded that just because we can now detect down to parts per trillion does not mean that carcinogens are not toxic at those levels, because of our general lack of understanding of mechanisms of carcinogenicity at the molecular level. He therefore argued that repealing Delaney would be bad science: high-dose animal studies can tell us that a chemical may cause cancer, but they cannot, without mechanistic information, tell us at what low-dose point carcinogenesis may be initiated in humans.[55]

This argument prompted an attack on high-dose animal studies by Warren Stickle, president of the Chemical Producers and Distributors Association, who asserted that a forty-pound child could eat 340 oranges every day for the rest of her life and not consume the amount of pesticides that it takes to cause health effects in lab rats.[56] Taking a step back, George Gray from the Harvard Center for Risk Analysis advocated approaching risk assessment from a total risk perspective. Gray pointed out that DDT, for example, was toxic to the environment but, as far as we know, not to people. Yet DDT was replaced by parathion, which is highly toxic to people, and agricultural workers in particular have high exposure potential to parathion.[57] Lynn Goldman, EPA's Assistant Administrator for Prevention, Pesticides, and Toxic Substances, added an additional layer of complexity by pointing out that we have no viable methodology for understanding the relationship between ecological impacts and risks to humans.[58]

The important point of this example for the nanotechnology case is that while risk debates can become quite elaborate and data intensive (i.e., calculating), they rarely are compelling when compared with the promised benefits of new products brought to society through global industrialization. Compare the above account to the following description by Clayton Teague of nanotechnology's promise.

> It is a truly transformational technology, promising widespread applications in many fields, ranging from energy and medicine to agriculture and manufacturing. As these applications move from the laboratory to practical use, nanotechnology has the potential to help strengthen the economy, protect homeland and national security, improve public health and the environment, and raise the quality of life for all people.[59]

Debates over interspecies extrapolation factors used in laboratory rat studies or measuring parts-per-trillion residues cannot compete with the obvious benefits Teague identifies in his statement. Global industrialism is set up to produce, and capitalism is designed to deliver, tangible products with clear benefit, while regulatory science can only make tentative statements about possible risks that are easily converted by proponents of products into *possibilities of no risk*. The hedge for capitalism, therefore, is that the generation of regulatory science information demonstrates to liberal democratic society that capitalism supports greater understanding of potential risks, recognizing that more understanding always will be needed—pushing actions on possible risks ever farther into the future—and that should a real debate actually arise on the risk of a beneficial process or product, the weakness of regulatory science can be exploited to turn a risk debate into a debate about the lack of evidence of risk. And as the regulatory science data are being generated and debate positions are being formed, through daily marketing and market penetration capitalism reinforces within society the benefits of its products.[60]

The dominance of the scientism/regulatory science story line within the nanotechnology risk discourse, enveloped in Weber's hard shell of capitalism that rests on Marcuse's one-dimensional rationality for modern industrialism, does not offer much room for optimism that resistance against current social forces would lead to enhancement of discourse around environmental protection. But how does our calculating democracy work with, or against, the power implications of global capitalism? This chapter concludes with considerations of such interactions in terms of their possible impact on power relations within the nanotechnology risk discourse.

CONCLUSIONS ON RISK, POWER, CALCULATION, AND CAPITALISM

With respect to nanotechnology, power over life and well-being will remain with science and technology unless there is some weakening in the dominant, co-dependent cultural traits of scientism and technological progressivism, manifest principally in the policy arena through regulatory science. Capitalism is not likely to be much help with this. But our calculating democracy might. In particular, using government institutions' need for transparency to objectify and justify their decisions will be avenues explored in chapter 6 in strategies for altering power relations within the nanotechnology risk discourse. Before going there, however, it will be useful to give a final consideration to how in the nanotechnology case, risk, calculating democracy, and global capitalism interact with one another to reinforce regulatory science's restrictive hold over how biopolitical considerations are addressed in discourse.

Jasanoff's concept of co-production, introduced in chapter 2, is an apt place to start when considering interactions between calculating governance and global capitalism within the regulatory science context. Regulatory science as a practice is defined by the needs of the state, and how the state addresses environmental issues is shaped by the capabilities and limitations of regulatory science.[61] It could be argued that science was born out of global capitalism's very origins, with much of the Enlightenment project driven by technological inventiveness stimulated by military and industrial endeavors, and that science still serves largely at the bidding of these pursuits.[62] That said, global capitalism must reach some accommodation with regulatory science if goods are to be traded within governance systems such as that employed by the OECD for the chemicals regime. At the same time, however, another co-production occurs: global capitalism shapes calculating liberal democracy, and liberal democracy attempts to exert discipline over global capitalism. Desrosières states succinctly the implications surrounding co-production between economy, state, and science:

> The question is therefore . . . to study how the tension between the claim to objectivity and universality, on the one hand, and the powerful conjunction with the world of action, on the other, is the source of the very dynamics of science, and of the transformation and retranslations of its cognitive schemes and technical instruments.[63]

In the nanotechnology case, regulatory science is retranslated within the chemicals regime to address particles produced at the nano scale. However, the world is a different place from when the chemicals regime originated. I would argue that a Risk Society, albeit a weak one, has grown up with expectations of quantitative characterizations of risk for chemicals that regulatory science may not be able to produce for nanoparticles, at least in any time soon with the same degree of confidence (low though it may be) that policy makers, industry, and other stakeholders accept for chemicals.

At the same time, "the world of action" in the form of global capitalism is putting pressure on liberal democratic regulatory regimes to accept nanotechnology products. As of 2010, EPA had received over one hundred premanufacturing notices indicating intent to produce specific types of nanomaterials, and also in 2010 EPA issued its first proposal to register a pesticide product containing nanoparticles.[64] Of the hundreds of products in the global marketplace that contain nanomaterials, some have gone through government regulatory processes and others have not. Many are sold solely on-line, with no clear indication of where they are manufactured or what types of nanomaterials, if any, the products actually contain.[65]

We have, therefore, a case with nanotechnology whereby global capitalism is using new methods such as e-commerce to market and distribute the products of new technologies on which we have little information about environmental impact. To maintain public trust in government's ability to protect citizens from harm, calculating liberal democracies have turned to regulatory science to provide data-driven, quantitative characterizations of safety and risk. Yet so far regulatory science has not proven up to the task. Nor do the enforcement capabilities of governments seem capable of even determining what nanotechnology-enabled products are already on the market, in need of regulatory testing, or are in violation of the law for not going through such testing.[66] In this instance, environmental governance appears to be breaking down. More likely, however, is that regulatory science has never really been fully up to the task of protecting democratic societies from environmental contaminants—the nanotechnology case only makes the problem more explicit.

The problem is largely one of biopolitical power. We know that global capitalism is a powerful force over life and death; we also know that regulatory science is a weak system that, by itself, will not provide evidence of risk compelling enough to counter capitalism's need to continual industrial development and product commercialization. In the middle is calculating democracy. Our need to order, number, and quantify is both a strength and a weakness of modern liberal democratic society: it enhances transparency, but simplifies debate; it acts as a disciplining force upon capitalism by forcing the generation of data, but the sheer volume of information it produces results in either important nuances being lost or having messages reduced to shallow outformations that impoverish debate and lend themselves to hijack by political and commercial interests. Nevertheless, democracies offer opportunities for discourse not available within nondemocratic regimes, and while calculation will continue to be prominent and powerful within democratic institutions, there may be opportunities to alter the biopolitical power dynamic through strategies and approaches that broaden debates on environmental risk. The following chapters move us down this path.

NOTES

1. Foucault, *History of Sexuality*, 100.
2. Agamben, 119.
3. Mills, 17.
4. Foucault, *History of Sexuality*, 143. Also see treatment of this issue by Mills, 21–22.
5. I consider "environmental risks" to be risks to humans, other individual species, and ecosystems.
6. Jürgen Habermas, *The Postnational Constellation: Political Essays* (Cambridge: The MIT Press, 2001), 134.
7. OECD, "OECD Programme on the Safety of Manufactured Nanomaterials 2009–2012: Operational Plans of the Projects," *Report to the Joint Meeting of the Chemicals Committee and the Working Party on Chemicals, Pesticides and Biotechnology*, OECD document ENV/JM/MONO(2010)11, 26 April 2010, http://www.oecd.org/officialdocuments/displaydocumentpdf?cote=ENV/JM/MONO%282010%2911&doclanguage=en.
8. See the ORD and OCSPP nanotechnology web sites: http://www.epa.gov/nanoscience and http://www.epa.gov/oppt/nano. ORD and OCSPP co-led the development of EPA's *Nanotechnology White Paper*, http://www.epa.gov/nanoscience/files/epa_nano_wp_2007.pdf.
9. Rachel Carson, *Silent Spring* (New York: Houghton Mifflin, 1962). In addition to creating general societal awareness of adverse environmental effects from pollution, Carson's assertion that the impacts of chemicals on certain individual species is a reflection of broader ecological degradation serves as the rationale for present-day ecological effects regulatory science testing approaches.
10. Jasanoff, *The Fifth Branch*, 95.
11. Sara Goodman, "Sen. Lautenberg Introduces Chemicals Reform Bill, Saying Current Regulation 'Is Broken,'" *New York Times*, 15 April 2010, www.nytimes.com.
12. US Senate, "A bill to amend the Toxic Substances Control Act to ensure that risks from chemicals are adequately understood and managed, and for other purposes," 5, http://lautenberg.senate.gov/assets/SCA2010.pdf.
13. For example, Vicki Colvin's statement that "the campaign against GMOs was successful despite the lack of sound scientific data demonstrating a threat to society." US Congress, House Science, 9 April 2003, 51.
14. Kleinman, *Science and Technology in Society*, 17.
15. Litfin, 38.
16. Even a non-anthropogenic source of potential impact such as radon gas is only considered an environmental risk when humans decide to locate structures at the site of radon emissions, thus making the risk the result of human action.
17. Silver Institute, Silver Nanotechnology Working Group, "Comments of the Silver Nanotechnology Working Group for Review by the FIFRA Scientific Advisory Panel; re: Re: FIFRA Scientific Advisory Panel; Notice of Public Meeting Docket ID: EPA–HQ–OPP–2009–0683," 15. www.regulations.gov and http://www.silverinstitute.org/images/stories/silver/PDF/epa_hq_opp2009.pdf.
18. Haraway, 47.
19. Haraway, 59 and 68.

20. Latour, *We Have Never Been Modern*, 32.
21. Merchant, Carolyn, *The Death of Nature: Women, Ecology, and the Scientific Revolution* (San Francisco: Harper, 1989 [preface published 1990]), 169.
22. Merchant, Carolyn, *The Death of Nature: Women, Ecology, and the Scientific* Revolution (San Francisco: Harper, 1989 [preface published 1990]), 171, 189.
23. Merchant, Carolyn, *The Death of Nature: Women, Ecology, and the Scientific Revolution* (San Francisco: Harper, 1989 [preface published 1990]), 228.
24. Merchant, Carolyn, *The Death of Nature: Women, Ecology, and the Scientific Revolution* (San Francisco: Harper, 1989 [preface published 1990]), xviii.
25. Porter, 215.
26. Nikolas Rose, The *Politics of Life Itself*, 3.
27. Ezrahi, 24.
28. www.bls.gov. Alain Desrosières makes this point: "Unemployment. . . . The very fact of equipping these objects with statistical tools in order to debate them—all this entered into a new language." Alain Desrosières, *The Politics of Large Numbers: A History of Statistical Reasoning*. Translated by Camille Naish (Cambridge: Harvard University Press, 1998), 199.
29. Vicki Colvin, "Regulation? Wait for Standardization, Commercialization," *The Environmental Forum*®, Reprinted by permission from the Environmental Law Institute, Washington, DC, March/April 2004, 35.
30. US Congress, House Science, 31 October 2007, 67.
31. Yaron Ezrahi, "Science and the Political Imagination," in Jasanoff, *States of Knowledge*, 261–262.
32. Porter, 196.
33. Ezrahi in Jasanoff, *States of Knowledge*, 254, 260.
34. ETC Group, "Nanotech News in Living Colour: An Update on White Papers, Red Flags, Green Goo, Grey Goo (and Red Herrings)," *ETC Group Communiqué*, issue 85, May/June 2004, http://www.etcgroup.org/upload/publication/95/01/livingcolorfinal.pdf.
35. Vicki Stone, et al., *Final Report: Engineered Nanoparticles—Review of Health & Environmental Safety*, European Joint Research Center, August 2009, http://nmi.jrc.ec.europa.eu/project/ENRHES.htm, 141.
36. US EPA, "Nanomaterial Case Studies: Nanoscale Titanium Dioxide in Water Treatment and in Topical Sunscreen (External Review Draft)," EPA/600/R-09/057, July 2009, www.epa.gov/nanoscience, 1–7.
37. Ezrahi, *The Descent of Icarus*, 41–42.
38. Ezrahi, *The Descent of Icarus*, 64.
39. ASTI Nanotechnology Research Institute, "China Is To Become More Globally Competitive in Nanotech Industry," *Asia Pacific Nanotech Weekly*, vol. 33, article 33, 2005, http://www.nanoworld.jp/apnw/articles/library3/pdf/3-33.pdf. Alexander Zaitchik, "Russia Pours Billions in Oil Profits Into Nanotech Race," *Wired*, 1 November 2007, http://www.wired.com/science/discoveries/news/2007/11/russian_nano.
40. Michael Hardt and Antonio Negri, *Empire* (Cambridge: Harvard University Press, 2000), 166.
41. Marcuse eloquently makes the argument that the nature of modern industrialism is totalitarian: "Thus emerges a pattern of *one-dimensional thought and behavior*

in which ideas, aspirations, and objectives that, by their content, transcend the established universe of discourse and action are either repelled or reduced to terms of this universe. They are redefined by the rationality of the given system and of its quantitative extension." Herbert Marcuse, *One-Dimensional Man* (Boston: Beacon Press, 1964), 12.

42. Herbert Marcuse, One-Dimensional Man, 37.

43. Society of Manufacturing Engineers, http://www.sme.org/cgi-bin/get-event.pl?--001624-000007-019902--SME-.

44. This is particularly important for businesses stung by popular backlashes against genetically modified crops.

45. Executive Office of the President, President's Council of Advisors on Science and Technology, *Report to the President and Congress on the Third Assessment of the National Nanotechnology Initiative*, 12 March 2010, 25.

46. Executive Office of the President, President's Council of Advisors on Science and Technology, *Report to the President and Congress on the Third Assessment of the National Nanotechnology Initiative*, 12 March 2010, 19.

47. Andrew Maynard, "Ten Things Everyone Should Know About Nanotechnology Safety."

48. US Congress, House Science and Technology, 16 April 2008, 94.

49. US Congress, House Science, 31 October 2007, 51–52.

50. David Baxter, "Nanotechnology: An Insurer's Perspective," *Safenano.org*, February 2008, http://www.safenano.org/NanoInsurancePerspective.aspx.

51. Sarewitz, 158.

52. Karl Popper (1902–1994) argued, like David Hume (1711–1776) before him, that deduction was the only valid approach to scientific explanation. Popper's basis for this position on *the problem of induction* was his belief that scientific laws are not supported by evidence, and that "there is no such thing as a logical method of having new ideas, or a logical reconstruction of this process." Rather, scientists use their imaginations to make conjectures about how the world works, and then set up experiments in an attempt to refute those conjectures. For Popper, induction has no place in scientific explanation. See Karl Popper, "The Problem of Induction," in Martin Curd and J.A. Cover, *Philosophy of Science: The Central Issues* (New York: W.W. Norton and Company, 1998), 429.

53. Bosso, *Pesticides and Politics* 97; and National Council for Science and the Environment, CRS Report for Congress, "The Delaney Clause Effects on Pesticide Policy," http://www.cnie.org/NLE/CRSreports/Pesticides/pest-1.cfm, 1. The 1958 Food Additives Amendment to the FFDCA Act requires FDA approval of the use of an additive prior to its inclusion in food. It also requires the manufacturer to provide proof that an additive is safe as used in the food. 21 U.S.C. §348(c)(3)(A). See also the web site of the FDA's Center for Food Safety and Applied Nutrition, http://www.cfsan.fda.gov/~lrd/foodaddi.html.

54. US Congress, House Committee on Commerce, Subcommittee on Health and the Environment, *Food Quality Protection Act of 1995: Hearings before the Subcommittee on Health and Environment of the Committee on Commerce*, 1.

55. US Congress, House Committee on Commerce, Subcommittee on Health and the Environment, *Food Quality Protection Act of 1995: Hearings before the Subcommittee on Health and Environment of the Committee on Commerce*, 92, 98.

56. US Congress, House Committee on Commerce, Subcommittee on Health and the Environment, *Food Quality Protection Act of 1995: Hearings before the Subcommittee on Health and Environment of the Committee on Commerce*, 119.

57. US Congress, House Committee on Commerce, Subcommittee on Health and the Environment, *Food Quality Protection Act of 1995: Hearings before the Subcommittee on Health and Environment of the Committee on Commerce*, 44.

58. US Congress, House Committee on Commerce, Subcommittee on Health and the Environment, *Food Quality Protection Act of 1995: Hearings before the Subcommittee on Health and Environment of the Committee on Commerce*, 31.

59. House Science, 17 November 2005, 21.

60. It has yet to be shown that the purported "no data, no market" ideal of the European REACH program will do much if anything to change this situation. At present, as many if not more nanotechnology products are being marketed and used in Europe as in the United States, with the same environmental data deficiencies as exist in the United States. For an overview of nanotechnology in Europe, see Landmark Europe, "Nanotechnology in Consumer Products," July 2009, http://www.landmarkeurope.eu/nanotech.html.

61. Jasanoff, *States of Knowledge*, 12.

62. Jardine, 6. Jardine's work focuses on how technology enabled scientific development. For discussion of how scientific discovery was stimulated by the needs of industrialization, see Peter Galison, *Einstein's Clocks, Poincare's Maps: Empires of Time* (New York: W. W. Norton & Company, 2003).

63. Desrosières, 7.

64. US EPA, "Control of Nanoscale Materials Under the Toxic Substances Control Act," http://www.epa.gov/oppt/nano; US EPA, "Proposed Decision Document for the Registration of HeiQ AGS-20, http://www.regulations.gov/search/Regs/home.html#documentDetail?R=0900006480b2f27e.

65. Project on Emerging Nanotechnologies, "Nanotechnology and Consumer Products," 18 August 2009, http://www.nanotechproject.org/process/assets/files/8278/pen_submission_cpsc.pdf, 2.

66. Project on Emerging Nanotechnologies, "Nanotechnology and Consumer Products," 18 August 2009, http://www.nanotechproject.org/process/assets/files/8278/pen_submission_cpsc.pdf, 2.

5

Implications for Environmental Risk Debates

In this chapter I describe how four findings from this study of the nanotechnology risk discourse have important implications for the possibility of broadening and enriching debates on environmental risks, in particular risks from emerging technologies. The four findings, drawn from the analyses in chapters 2, 3, and 4, are:

- Scientism, manifest in regulatory science, is a story line that runs throughout the nanotechnology risk discourse.
- Language describing distinctions between natural and human-created objects causes uncertainty and instability within discourse about nanotechnology risks.
- Overlap statements blur boundaries between discursive categories.
- Influential discourse participants have a common set of skills and attributes.

I begin this chapter with a brief restatement of my argument on the narrowness of risk debates. In support of this argument, I look at "other voices" that, in their absence from current nanotechnology risk debates, demonstrate the potential for adding new perspectives to policy discourse. I follow with discussion of the four study findings. Building on chapter 4's evaluation of biopower, I analyze how power relations operate within discourse with respect to each of the four study findings. Consistent with Foucault's observation of the imbrication of power and knowledge within discourse,[1] this treatment of power relations is important for laying the groundwork for the policy recommendations in chapter 6. I also evaluate what the study findings imply for two tools I propose for use in enhancing environmental risk debates: STS scholarship and targeted policy initiatives. I conclude this chapter with my perspective on what the study findings say about the current state of the nanotechnology risk discourse, and what the present state of the discourse implies for how the potential environmental risks from emerging technologies may be addressed in the public policy arena.

RESTATING THE ARGUMENT

My main contention, which also has been my motivation for conducting this study, is that environmental risk is described too narrowly in American public policy. This narrowness results from the policy arena's technocratic reliance on quantitative risk assessment. Characterizations of risk are constructed, for the most part, from experimental or modeled data on the potential for a pollutant or set of pollutants to cause an adverse effect on an individual biological system, a population, or an integrated ecological system. The primacy of scientism as a story line running through environmental discourses, manifest in the dominance of regulatory science within risk discourse, is the principal cause of the policy arena's narrow view of risk. I see this narrowness as a problem specifically for considering the environmental impact of emerging or future technologies, as well as more generally for environmental protection, and more widely still for the conduct of public policy within liberal democracy. Yet I see the problem as tractable, since means exist for broadening and enriching environmental risk discourse. My argument, therefore, states a problem that is solvable, or at least can be made less of a problem. The key findings analyzed throughout this chapter point to some of the challenges to be faced in addressing the risk debate problem; they also offer insights into how such challenges may be confronted and overcome.

OTHER VOICES, PARALLEL DISCOURSE

It is true that this book is concerned primarily with broadening and enriching environmental policy discourse by bringing additional perspectives to the discourse surrounding nanotechnology. Such inclusion is important, because it is within environmental policy discourse that governance of nanomaterials now occurs and is likely to continue into the foreseeable future. Yet there are other voices that despite attempts at inclusion may for some time continue to remain outside this discourse, even though they are highly relevant to the environmental impacts of nanotechnology. These voices represent discourse occurring parallel with, and rarely if ever intersecting, the scientism-driven environmental policy discourse. Such voices ought to be acknowledged. This section sets out to give such acknowledgment as well as to show how alternative framings of the issue of technology's environmental impacts eventually may open up paths of entry for such voices into the policy arena.

Before doing so, I wish to address the question of why one should give attention to discourse that is not presently, and may never be, influential to policy making. This is a fair and practical question, particularly for a book that above all seeks to stimulate policy-relevant action toward enhancing and advancing risk debates. The answer is, quite simply, that life is bigger than the public policy arena. Even in a calculating democracy such as ours that is tightly locked into a system of global capitalism, narratives are constructed and conveyed through society in speech and actions that have neither resonance in the halls of Congress nor repercussions on the global economy. Yet they are part of the human yearning for understanding and therefore matter.

Less simply but more directly to this book's policy-relevant focus, a parallel discourse at some point will intersect, sometimes will merge with, and on rare instances will become a replacement for, the dominant narrative within society. An example of such intersection is the sporadic impingement into western capitalism of what Dryzek labels the green romanticism discourse (which includes, among other areas, the deep ecology movement and cultural ecofeminism).[2] Perhaps the most-recent high-profile meeting of these parallel discourses was manifest in the 1999 protests in Seattle over the meeting of the World Trade Organization. In Seattle, environmental advocates joined with labor organizations and others to challenge basic premises concerning global capitalism's contribution to human and environmental well-being.[3] The long-term implications of this point-in-time physical manifestation of an ideological clash are probably not yet knowable, but continued references to the event more than a decade later suggest that in some way discourse was altered by the event.

Public reaction to the first photographs of Earth taken from outer space signaled an instance of a heretofore fringe Malthusian discourse merging with the dominant societal view of limitless resources. The 1972 publication of *Limits to Growth* pulled neo-Malthusian perspectives into the western vernacular, and Earth as a "big blue marble" became a standard image in mainstream environmental policy discourse.[4] Thus entered Earth as a discrete entity into environmental discourse. However, Earth as *finite* remains in dispute, and thus neo-Malthusian perspectives have entered but by no means have subsumed the discourse. Indeed, the extent to which, or whether at all, the Earth's resources are finite remains contested territory. Nevertheless, the idea of resource conservation is now firmly embedded in environmental policy discourse, less than fifty years after its emergence into social awareness.

A rare instance of a minor, parallel discourse replacing the dominant discourse can be found in the change in treatment of the idea of nature itself over the past one hundred years. As has been pointed out by feminist writers such as Carolyn Merchant, much of the Enlightenment project has been dedicated to using science and technology to bring nature under human domination. Yet Merchant also reminds us, through examples such as opposition in the seventeenth century to the draining of the English fens, that even as Enlightenment views gained force in western society, there existed parallel (albeit generally unsuccessful) discourse around the value of balanced coexistence with the natural world.[5] From the late-nineteenth century activism of John Muir to the establishing of the EPA in 1970, a profound shift had occurred in discourse around nature. In language, if not always in practice, nature now is held within western liberal democratic discourse as something to be preserved and nurtured, rather than exploited. This is a dramatic reversal that, with so much to be justifiably concerned with over the state of the environment, is easy to forget.

It is clear, then, that softer voices can intersect, merge with, or replace those voices that dominate policy debates. Yet rarely do they do so, and this is a concern for the healthy development of liberal democratic society. The difficulty in bringing minority voices into policy-relevant discourse is an issue because of problems associated with tolerance, and a type of black boxing that unfortunately, and to the detriment of democracy, can happen to parallel, minority discourse.

In democratic society, perspectives that do not enter policy debates can be said to be *tolerated*: room is made for them within society, but the space these tolerated perspectives are given is located outside the political space in which power is exercised in language and practice. As articulated by political scientist Wendy Brown, issues are depoliticized as they become subjects of tolerance: they acquire an "ontological naturalness" that makes them an essential part of our culture, and this is not necessarily a good thing.[6] They are *essential* in that they are taken as they are, as givens, without any history or power associated with them. Debate on their history, meaning, or even social relevance is obviated through depoliticization. Once an issue is depoliticized, we no longer have to subject it to public debate—we just have to live with, or *tolerate,* it.

One might say that these tolerated perspectives are "black boxed"—not in Latour and Woolgar's sense of their becoming facts no longer subject to debate, but rather that they are set aside as discursive oddities that do not merit debate.[7] In society, depoliticizing an issue through tolerance can shift public discourse away from understanding the nature of the issue itself, and toward how society is going to tolerate the issue. Racial tolerance, for instance, is a longstanding national issue for America. Once we depoliticized race by making it a subject of tolerance, we presumably obviated the need to publicly discuss the nature of race. For example, instead of discussing whether race is truly a significant or even legitimate human identifier, or whether it perhaps is nothing more than a socially constructed concept with no more biological relevance than, say, hair color or left- or right-handedness, we solidify, or *make essential,* race as a significant national issue and continue the discussion on that basis.

Brown's notion of ontological naturalness resulting from depoliticization has significant power implications. Deciding to black-box is a decision to end inquiry, be it scientific inquiry or social inquiry and dialogue. There always are interests that are keen to end debate, and others that want to keep the discussion going. If the dominant forces in democratic liberalism (even advanced democratic liberalism) can use tolerance to control the depth and direction of national-issues dialogues in a manner that undermines the interests of those in subordinate power positions, then tolerance is indeed a dangerous political tool.

This raises the question of the place of the advanced liberal democratic state institutions in the tolerance-as-depoliticization-tool power equation. It may be too much to ask of the state to proactively reopen debate that has been suppressed through depoliticization's naturalization process: those who manage the state, both elected officials and career civil servants, are too much a part of the dominant social interests that boxed up tolerated discourse in the first place. Rather, what we may rightly ask of the state is that, when a critical national event reopens debate—for example, as Barack Obama's being elected president of the United States has offered an opportunity to reopen national dialogue on race—state institutions do not unwittingly stifle discourse by falling into the tolerance trap. Black boxes are seldom reopened, and we should take care that when they are, we are not too quick to seal them back up.

This discussion of tolerance has direct implications for debates on the environmental implications of nanotechnology. As I have argued throughout this book, the regulatory science view of risk—maintained through discourse dominated by a scientism narrative—controls debates not only on nanotechnology but generally regarding the environmental implications of technology. Other ways of viewing the relationship between human activity and the rest of the biosphere, such as Gaia theory, are tolerated (journal articles and books on Gaia abound) but as with other tolerated discourse are depoliticized and thus not empowered to participate in policy debate. In congressional hearings, government agencies, companies, think tanks, and so on, it is politically correct to acknowledge, to *tolerate* the notion, that all we label "the environment" is an integrated, interdependent system. After such polite acknowledgment, discourse quickly moves back to talking about using regulatory science to address risk.

What then is the answer? To let parallel, minor, tolerated discourse run its course outside the policy arena, with the possibility that in the next year, decade, or century it may intersect, merge with, or replace the dominant scientism narrative? We could do so, but odds are no such intersection, merging, or replacement would happen, at least not in our lifetimes. I believe one answer lies in identifying ways to re-politicize narratives that have been packed away in boxes of tolerance.

Re-politicize is an important term, because like all tolerated subjects, alternative views of environmental impact once were political, or at least the ideas behind those views were, but over time were stripped of power as institutions of governmentality worked to develop and advance western environmental discourse. The evolving location of Gaia theory in environmental policy discourse is an illustrative example of how discourse may become repoliticized and enter policy debate.

As described by Eileen Crist and H. Bruce Rinker, adherents to Gaia theory see Earth's physical and chemical conditions as inseparable from the existence of life on Earth.[8] In other words, to what extent is Earth, as a single entity, "alive"? Gaia theory encompasses a diverse range of perspectives, from the "strong Gaia" view that Earth's biota regulate conditions in the biosphere to keep it optimal for habitation, to the "weak Gaia" position that biota affect the global environment. In between these two poles is the current dominant Gaian view that the biotic and inorganic aspects of the world have coevolved into a tightly integrated global system of interactions and feedbacks that work to optimize the conditions for life on Earth. A key research question across the Gaian spectrum, therefore, is to what extent are Earth's living and nonliving entities coevolving in an integrated way that serves to regulate the planet's conditions.[9]

Gaia was not well received by the environmental science community when it was first discussed in the 1970s. Part of this poor reception was a negative reaction to strong Gaian implications of a *telos* to the behavior of the biosphere, an idea for which environmental scientists were quick to point out there was little scientific support.[10] Also, Gaia theory in general runs counter to the reductionist direction that science has taken since the Enlightenment. However, within environmental science, systems thinking has developed considerably in recent decades, and such thinking is not entirely inconsistent (though neither entirely consistent) with Gaia theory.

It is in the political and public policy domains where thinking remains at best tolerant of, but more often hostile to, even systems thinking, let alone Gaia theory. As Karen Litfin observes, Gaia theory's assertion of interdependence among all living (and some nonliving) systems, including humans, runs counter to western society's political and psychological beliefs in human independence and autonomy.[11] Policies are formulated and legislation is enacted to mandate certain institutions to address specific issues. Environmental problems are not considered within the context of their relationship to the other interactions occurring within the biosphere that, according to Gaia theory, sustain and regulate life on Earth.

Policy makers' resistance to repoliticizing Gaian thinking, or anything close to it, is illustrated in the inability of the political arena, particularly in the United States, to manage a rational discussion of global climate change. While much of this resistance is due to the integral relationship between anthropogenic greenhouse gas emissions and global capitalism, some of it also results from the dominant scientism discourse—in particular risk-based regulatory science—discussed throughout this book. Environmental systems theory, developed within the scientism discourse, tends to be constructed within policy discourse as sets of hard connections that are understandable, and through the modeling of which estimates of causality can be made of the impact of human changes to the system and the system's response. As Crist points out, this belief in our ability to "model" the environment engenders confidence among policy makers in our capability to apply scientific or technological measures to mitigate our environmentally unsustainable behavior.[12] Geoengineering approaches to addressing climate change, such as releasing aerosols or nano-scale mirrors into the atmosphere, are examples of twenty-first century manifestations of the Enlightenment belief in human capacity to dominate nature.

Gaia theory would seem to challenge not only technology fixes such as geoengineering, but also systems theory presumptions that the vast network of interactions within the biosphere can be understood and possibly managed, as one might regulate some mechanical system. Instead, Gaia theory suggests that the forces that regulate the biosphere are much more powerful than any forces that humanity could muster, and therefore the only hope for continued flourishing of the human species is to realign itself into a hospitable relationship with the rest of the biosphere to which it is inextricably linked. Litfin makes the point very clearly: "From a Gaian perspective, we [humans] are far more expendable than bacteria."[13]

The hesitancy of policy makers to latch onto geoengineering schemes, and the inability of regulatory science to bring order to climate change debate, suggests that in some small way Gaia thinking, even if not labeled such, is becoming repoliticized. The idea of a living Earth is being challenged, explored, and analyzed as the potential implications of climate change are discussed and federal agencies explore mitigation and adaptation strategies. Whether or not human activity is harming the Earth is being hotly debated and positions at the poles of the debate are hardening. However, whether or not those participating in the debate believe there is a connection between human activity and the rest of the biosphere, the issue of living Earth is not being ignored.

What, then, does this mean for the nanotechnology environmental debates? It means that alternative framings of issues surrounding nanotechnology and the environment—that is, alternatives to the dominant regulatory science/risk framing—may find entry into the political arena. The Gaia case may point the way to achieving such politicization: by exploiting the uncertainty inherent in regulatory science.

Chapter 2 described many of the uncertainties that are inherent in identifying, characterizing, and describing risks from manufactured nanomaterials. It is the job of those conducting regulatory science to attempt to reduce these uncertainties. However, risk assessment issues are framed in terms of reducing the uncertainties in risk estimates to a specific substance or set of substances. So long as environmental impacts are framed in terms of discrete risks, the position of regulatory science remains dominant within environmental policy discourse. However, a re-framing away from risk, to improving the human and environmental condition in the broadest sense, has the potential to extend political power in other directions. Such re-framing will be a topic of discussion in chapter 6's presentation of targeted policy initiatives for enhancing discourse on the environmental implications of nanotechnology.

KEY FINDINGS

Table 5.1 summarizes what I consider to be the principal implications of this work's findings for *power relations*—since, per Foucault, power relations are the main determinant of knowledge construction within discourse—and *STS scholarship* and *targeted policy initiatives*, which are the two tools I put forward as useful for enhancing environmental risk debates. The implications discussed in this chapter will, in my view, shape the direction of public policy discourse on risks arising from the introduction into society of new technologies.

Regarding the implications of the key findings for power relations, chapter 4 develops my contention that social forces, in particular calculating governance and global capitalism, exert biopolitical power within discourse in ways that limit the range of voices heard in the nanotechnology risk debates. This contention, and the support for that contention, is grounded in the idea that knowledge is constructed through the power relations that operate within discourse. As Foucault states, "One would have to speak of *bio-power* to designate what brought life and its mechanisms into the realm of explicit calculations and made knowledge-power an agent of transformation of human life."[14] The centrality of biopolitical power to the social construction of risk is key to understanding the treatment of nanotechnology in discourse.

Table 5.1. Summary of Implications of Key Findings on Power Relations, STS Scholarship, and Targeted Policy Initiatives

Key Finding *Implications of Finding For:*	*Scientism/Regulatory Science Runs Through Discourse as a Story Line*	*Natural versus Human-Created Distinctions Create Uncertainty and Instability*	*Overlap Statements Blur Boundaries*	*Influential Discourse Participants Have Skills and Attributes*
Power Relations	Focus is on who defines risks, who decides what's risky, who addresses risks. Risk defined by toxicological endpoints. Government will decide nanomaterial risk. Power is retained in those who control regulatory science.	Unstable situation for chemical regime regulators. Biopower implications of uncertainty in data needs. If public sees nanoparticles as natural, may cede power to industry; if seen as human products, may cede power to government.	Reinforced privilege of regulatory science. Diffusion of power broadly across discourse and throughout debate.	Concentration of power in who defines what risks exist and how they should be addressed. Opportunities exist to create new influential participants in debates.
STS Scholarship	Using discourse analysis to define boundaries between science and technology would be useful.	The language used to describe environmental stressors may reveal power relations in technology debates.	Need methods for combining discourse analysis and ethnography to analyze cultural implications of language.	Opportunities to use ethnography to understand how symbols influence social behavior.
Targeted Policy Initiatives	Lack of institutional capacity to meet data needs could lead to diminished public trust of institutions. Regulatory science dilemma: wanting to anticipate risks but needing data to define risks. Regulatory science dominance will continue to narrow policy debate.	Uncertainty will shape how benefits and costs/risks are characterized in environmental policies and decisions, as well as what policy tools are available.	Unsettled thinking and positions on nanotechnology risks provides an opportunity for new voices in debates.	Understanding key skills and attributes presents opportunities to develop and bring new voices into debates.

With respect to the implications of the key findings for STS scholarship, I contend that STS thinking is underutilized in the grappling over risk issues that are raised in the policy arena. Largely because of federal funding under the NNI, some STS research has been conducted for nanotechnology. However, little of this research takes on risk issues within the context of public policy and regulatory science.[15] Beyond nanotechnology, but highly relevant to it, are rich STS-related literatures that touch on the issues raised by this study's findings. As mentioned in chapters 1 and 2, scholars such as Kleinman and Jasanoff have addressed scientism and regulatory science, respectively; Haraway, Latour, and Merchant have written on the nature versus human distinction; Foucault, Wetherell, Dryzek, Mills, and numerous others have explored the power issues revealed in what I have called discourse overlap statements; and Hajer, Halfon, and Litfin are three scholars cited in this work who have discussed how skilled individuals can influence political strategies within discourse.

FINDING 1. SCIENTISM, MANIFEST IN REGULATORY SCIENCE, IS A STORY LINE THAT RUNS THROUGHOUT THE NANOTECHNOLOGY RISK DISCOURSE

The 665 statements analyzed as research for this book, and discussed in chapters 3 and 4, do not question the desirability, let alone the legitimacy, of framing risk concerns in terms of regulatory science. Statements employing risk society language reflect a worry that we have insufficient scientific data to understand what types of environmental impacts nanotechnology will cause. Technological progressivism language expresses confidence that data will demonstrate that nanotechnology's benefits outweigh its risks, and that we can handle the risks. For statements I have organized within the administrative pragmatism discursive category, generating and applying scientific data to calculate risk probabilities and base decisions on those probabilities is the logical, responsible, transparent, and defensible path forward. In all cases, the implication is that we need, and should redouble our efforts and resources to obtain, more scientific data and act (or not act) in a manner consistent with how regulatory science characterizes risk through its assessment of the data.

Looking across all the statements, I come away from my evaluation of nanotechnology risk language with the conclusion that all nine speakers of those statements assume that regulatory science is the best approach for showing us the way to successful and environmentally protective development, deployment, and use of nanotechnology. So firmly embraced is this assumption that absent from the entire debate is any notion of reflexivity regarding science. Scientism, in the form of regulatory science, is not just accepted: it defines the limits of discussion.

This may seem astounding, given regulatory science's poor track record in resolving environmental risk controversies. Yet for those heavily engaged in policy discussions, there appears to exist either high confidence in, strong loyalty to, or no recognition of an alternative to science as the privileged arbiter of nanotechnology's value to and impact on society. Scientism's grip on environmental policy discourse—as evidenced by the prominence of regulatory science for decades in the chemicals regime as well as in today's debates on nanotechnology—suggests that environmental policy's science-dominated narrative will carry forward into discussions and evaluation of the impacts of the many new technological advances society likely will face in the coming decades.

Power Relations

Jasanoff labels regulatory science "science of the people," in the sense that bringing science into public policy decisions makes science, through liberal democracy's transparent administrative procedures, accessible to public view and interaction in a way that does not exist with traditional research science.[16] Making science public, Jasanoff argues, gives power to regulators to interpret and apply science, as well as gives power to the courts to uphold or reject the legal defensibility of regulators' use of science in administering environmental statutes.[17] The power wielded by regulators, the courts, industry, and others involved in environmental policy is biopower. In the nanotechnology case, scientism's bounding of the policy debate, and regulatory science's control over the debate itself, has biopolitical implications.

The principal biopolitical implications of regulatory science's dominance over the nanotechnology risk debate center around who defines risks, who decides what is risky, and how any identified risks are addressed. As discussed in chapter 2's description of regulatory science, each of these three questions is addressed through interactions between people and institutions across the American governmentality network.

In nanotechnology risk debates, power over life and death is defined in terms of regulatory science endpoints. The definition of what defines an environmental risk comes out of long-standing scientific practices, administrative decisions, case law, legislative mandates, and public demand for action on certain health issues, such as cancer. It is clear from the discourse that the dominance of regulatory science means that risks for nanomaterials will continue to be defined in terms of toxicological endpoints. For humans, this would include cancer, reproductive and developmental toxicity, and neurological effects. Many of the same effects would be evaluated for other living things but would focus on, among other things, impacts on wildlife populations and movement of nanomaterials through the food web. There is no center of power outside the governmentality system, which is heavily invested in this definition of risk, to change the definition. Based on the statements evaluated for this study, there is no indication of likelihood for a change in the power dynamic that has defined what risk means for nanotechnology policy.

The strength of scientism and regulatory science means that government—specifically, government scientists and science managers—ultimately will decide what nanomaterials end up being considered risky. I say "ultimately" because, along the tortuous route toward an official risk determination, scientists from industry, academia, environmental NGOs, and other participants in regulatory science will add their language—their statements of "facts"—to the mound of data upon which at some point will rest the regulatory science position on a particular nanomaterial. Each new nanomaterial study will create the temptation for those who seek to influence risk construction to put the nanomaterial into either a *risky* or *not risky* black box. Government agencies will make pronouncements on risk and make decisions consistent with such pronouncements, but typically that is not the end of the debate, as interests turn to the courts or Congress to argue that regulatory science was misused or ignored in the regulatory determination. Of course, the courts and Congress too are government, and therefore biopolitical power remains largely with the government institutions.

It is true that many statements in the nanotechnology debates express concern that the public will come to see nanotechnology as risky and reject it. However, the remedy offered by those making the statements is for government, and in some cases industry, to support more regulatory science. Therefore, irrespective of what potential biopolitical power the public may have regarding nanotechnology's future, presently within the nanotechnology risk discourse those in control of regulatory science wield the power.

Once regulatory science constructs a characterization of risk, it hands back over to science and technology the responsibility for addressing the risks that science and technology created.[18] On very rare occasions government requires the removal of a substance from the marketplace. Usually, government requires of industry either engineering or other technological controls to reduce exposure of the substance to humans or the environment, or prohibits specific uses of the substance based on scientific calculations of how much reduction must be achieved to lower exposure to levels deemed by regulatory science to be acceptable. Rarely is the social desirability of the substance itself brought into discussions of preventing or managing risks, since social factors, particularly if they are not amenable to quantified expressions of risk or benefit, do not fit well into calculation and regulatory science approaches.

Scientism and Regulatory Science: STS Boundary Drawing

Kleinman identifies scientism and technological progressivism as two discourses being conducted simultaneously in western society.[19] Jasanoff concludes that, to strengthen the defensibility of their decisions, governments are willing to trade off open, democratic policy making for decision-making approaches that emphasize expert knowledge and focus on narrow regulatory issues.[20] The pervasiveness of regulatory science across the statements analyzed in my study, which were statements oriented toward the potential environmental risks from the products of a platform technology, suggests a line of STS inquiry that can bridge the kinds of issues raised by Kleinman and Jasanoff: science-technology boundary issues as they relate to environmental risk.

Gieryn's account of John Tyndall's "double-boundary" effort to demarcate science from both religion and engineering may be a useful starting point for exploring environmental risk science-technology boundaries. Gieryn describes how in Victorian England, Tyndall used cultural tactics to achieve the economic goals of acquiring more resources for science.[21] Those tactics included describing how the technological achievements of the time, such as the steam engine, could not have been attained without the underpinnings of scientific knowledge, thus making the case that science was not only of practical relevance but also was an enterprise distinct from engineering.[22] There are similarities to the Tyndall case in how nanotechnology risk discourse relates to regulatory science, and these similarities raise opportunities for STS scholarship.

Just as Tyndall attempted to draw a boundary between science and technology, yet simultaneously demonstrate the latter's dependence on the former, so too does the regulatory science community engage in complex boundary drawing.[23] Scientism and technological progressivism both are gathered within the regulatory science boundary in two ways. First, they are joined through the belief that modern tools, such as toxicogenomics and bioinformatics, can reduce uncertainty in risk estimates.[24] Second, they come together under the assumption that regulatory science can identify and guide the management or prevention of risks from the introduction of technologies into society. As Jasanoff observes, there is also boundary drawing by both traditional hypothesis-driven science and regulatory science in terms of science's authority to establish knowledge claims about environmental risk.[25] To add to the complexity, there is even boundary drawing between scientific disciplines within regulatory science. For instance, in environmental science, health and ecological effects research often is held distinct from research on exposure, even though behaviors such as chemical bioavailability overlap these disciplinary lines.

The STS scholarship opportunity that the scientism/regulatory science finding for nanotechnology presents is to explore how the regulatory

Power Relations

The uncertainty and instability created by the distinction between nanoparticles as natural versus being human-created substances has biopolitical implications in two areas, one relatively narrow and the other quite broad. The narrower area resides within the existing chemical regime; the other area is the broader societal relation to nanotechnology as a possible creator of stressors on the environment.

Government's power to require, collect, and analyze regulatory science data defines, at least in the environmental policy arena, liberal democracy's ability to protect the public from harm. The uncertainty over whether manufactured nanomaterials are entirely new or are merely manipulations of natural substances creates an unstable situation for environmental regulators who are steeped in the practices of managing industrial chemicals. New substances under US environmental laws (such as the Toxic Substances Control act, or TSCA; and the Federal Insecticide, Fungicide, and Rodenticide Act, or FIFRA) may trigger government requirements for new regulatory science data. Variations of existing substances in some cases will be assumed to be covered under existing data already submitted for the substance. Therefore, the uncertainty over nanomaterials' place in the natural-human created/existing-new continuum destabilizes discourse over what regulatory science evidence will be required to determine the risk of nanomaterials. If regulatory science is society's accepted method of determining risks and protecting the public from environmental pollutants, how protected does society feel if government cannot decide what regulatory science information is needed to make decisions regarding the risk of nanomaterials?

At the dawn of the Enlightenment, nature was a scary place that humans sought to control in part through technology.[33] The axe, plow, and gun tamed the wilderness; electric light removed the limitations of darkness. As we enter the twenty-first century, this relation is mostly reversed: we seek refuge in parks, consider eating organically grown food to be a healthful practice, and expect government institutions to protect us from technology's harmful by-products. If the nanotechnology debate leads to a public perception that nanomaterials are, for the most part, natural substances, then society may be more likely to cede biopolitical power to industry in the form of lessened demand for government oversight of the materials' production and introduction into society. Indeed, if nanomaterials were to be seen as natural then there would be no notion of *introduction*—they would be considered to always have been with us. However, biopolitical power shifts back to government if nanoparticles are perceived as wholly human-created and unnatural. Like the European perception of genetically modified foods, they would be viewed as foreign substances—and as foreign potentially dangerous—and the expectation will be that government will play the role of protector.[34] Until language in policy debates settles on a position on where nanoparticles reside along the natural/human-created spectrum, there likely will continue to be fluctuations in power dynamics between industry, government, and the public over who controls the definition and articulation of nanotechnology risks.

Natural versus Human: Stability within Discourse

There has been much STS scholarship on distinctions between the natural and the human, and I touch on some of it throughout this work. Where I believe a new STS research opportunity may exist is in understanding what causes instability in discourse, and what, if anything, may lead discourse to settle at least for a time into stability. Foucault speaks of the need to look at a discourse as a "multiplicity of discursive elements" that operates within a "complex and unstable process" of interaction with power.[35] If we add to this the observation by Mills that discourses are "the site of constant contestation of meaning," then we have the beginnings of a framework within which to locate the sources of uncertainty and instability in the environmental protection episteme.[36]

For environmental policy-oriented STS scholarship, I would suggest that examination of environmental stressors—be they chemicals, invasive species, changes in climate, or nanoscale particles—in terms of the language used to describe such stressors' natural and anthropogenic origins and attributes, could reveal information about what makes power relations complex and dynamic when policy debates turn to the impact of technology on the environment. In the nanotechnology case, for example, what language is used to describe the relationship between the nanoscale particle and the human who creates the particle? That is, the relationship between the technology and the technologist. What political, economic, and other social forces act on that relationship? A case study approach that employs discourse analysis could prove valuable for exploring dynamics broadly within the environmental protection episteme, or more specifically within risk discourse.

The Nature-Human Distinction: Seeking Stability in Calculation

Earlier in this chapter I discussed how the power-relations implications of this key finding may impact how agencies go about collecting and using regulatory science data. The broader policy question I examine here relates to calculated governance. To make this discussion useful for the proposal put forth in chapter 6, I focus on calculated governance as it applies to public acceptance of government's articulation of nanotechnology risk issues.

Uncertainty over whether nanoparticles are natural or of human creation will impact how benefits and risks are framed within the nanotechnology risk discourse. This is a crucial issue because a central belief often articulated through administrative pragmatism language is that benefits and risks should be measured and calculated, and weighed against one another to inform decisions and actions on risks. Cass Sunstein, a proponent of using cost-benefit analysis to inform environmental decision making, argues for a number of propositions to make such analysis useful in capturing the impacts of regulatory decisions. One of those propositions is that cost-benefit analysis should include broad qualitative descriptions of the potential impacts of action and inaction on a risk issue, supplemented to the extent possible with quantitative analyses.[37] It is in such qualitative analysis that the uncertainty and instability of the natural-human distinction will become important.

By and large, US environmental statutes are set up to address risks from stressors of anthropogenic origin. The 2007 Supreme Court decision that greenhouse gases are air pollutants covered by the Clean Air Act, together with the EPA's 2009 finding that greenhouse gases endanger public health and well-being, exemplifies this tie between human-caused environmental impact and what are accepted as the legitimate boundaries for regulatory activity and the application of policy tools.[38] These boundaries also define the parameters that government would find acceptable for the kind of qualitative analysis advocated by Sunstein. Therefore, how closely nanoparticles or specific aspects of particles are tied to anthropogenic activity, as opposed to being seen as merely a repackaging of naturally occurring substances, may impact what is included—and perhaps more important, what is omitted—from how nanomaterial benefits and costs are described.

The statements analyzed in this study suggest that nanotechnology policy debates are unsettled on the nature of nanoparticles. In a 2009 meeting of EPA's FIFRA Scientific Advisory Panel to discuss the state of the science and regulatory data requirements for antimicrobial products containing nanoscale silver as the active ingredient, there was clear division among the interest groups who made presentations to the panel. Business interests argued that new products containing nanoscale silver particles were, from an environmental and public health perspective, the same silver that has been used for millennia for its antimicrobial action. On the other side of the debate were environmental NGOs, who argued that because the new products contained particles intentionally engineered at the nanoscale, with new properties that make it worthwhile for industry to incur the cost of manufacturing them and incorporating them into products, there is reason to suspect that those same new beneficial properties may cause new toxicological effects of concern.[39]

The silver nanoparticle case, while challenging, is simple compared to what regulatory agencies are likely to face in the coming years. Going beyond the study of particle interactions at the nano-bio interface,[40] biologists and material scientists are developing approaches to marry nanotechnology with synthetic biology.[41] If language and behaviors within discourse presently are uncertain and unstable because of ambiguity over where nanoparticles reside on the natural-human continuum, they likely will become even more so as interactions between nanoparticles and biological material increase in complexity. If the silver nanoparticle case is any indication, such developments will have ramifications for environmental and regulatory policy making. If such uncertainty and instability are manifest in indecision about the application of environmental statutes, this will have implications for the application of policy initiatives, which to be accepted by governance institutions must be viewed as consistent with the accepted interpretations of the boundaries of legal mandates.

FINDING 3. OVERLAP STATEMENTS BLUR BOUNDARIES BETWEEN DISCURSIVE CATEGORIES

In chapter 3, I discuss how overlap statements demonstrate the power of the scientism story line within the nanotechnology risk discourse, and how overlap statements join power to other important themes within the three discursive categories. In my analysis of nanotechnology risk language, I find that the overlap statements change the overall perspective on risk that one typically would expect to be conveyed by language employed within that category (that is, as compared to the perspective one would gain by stripping the overlap statements out of the category). For example, when within the risk society discursive category language is introduced that is supportive of administrative pragmatism, what occurs is that risk society's reflexive modernization—typically the dominant voice in the risk society—runs up against the technocratic analytical framework of regulatory science. That is, the mirror that risk society would endeavor to hold up to modernity to reflect the risky reverse image of technology's benefits is made opaque by the scientific uncertainty, or underdetermination, inherent in the calculated, evidentiary standard by which administrative pragmatism requires technology to be evaluated.

One could argue that the impacts of overlap statements on discursive categories could be avoided by defining or setting the boundaries of discursive categories differently. I would agree that this probably could be done. But this really is not the point. What is important is that the blurring of boundaries is yet another nuance to a risk discourse that is imbued with scientism, and in particular with regulatory science. Within nanotechnology risk discourse, one can employ any number of linguistic strategies—the three discursive categories I constructed represent just one approach for considering such strategies. Yet irrespective of which type of language one uses in a given situation, strategies employed within discourse will be shaped by a scientism narrative that carries with it power implications that go beyond use of language and perspectives on risk.

Power Relations

Philosophers such as Susan Haack and Philip Kitcher argue that society ought to consider science to be epistemologically *distinguished*, but not privileged.[42] Yet the overlap statements in my discourse analysis seem to reinforce scientific privilege, specifically the privilege of regulatory science in shaping environmental policy. For instance, the boundaries around the language I have placed within the risk society category are blurred by statements that reflexively question the net societal benefits of nanotechnology but call for regulatory science to address the risk side of the benefit-risk equation. This particular boundary blurring acknowledges that the very power reflexive modernity brings to risk discourse—the power to question the benefits of science and technology—is pulled by regulatory science into the narrative of scientism. Likewise, technological progressivist language that confers *a priori* justification of nanotechnology's beneficial place in society is weakened if that same language seeks to bolster its benefits argument by citing a lack of regulatory science evidence of environmental risk from nanoparticles. In this case, the scientism narrative draws power away from the discursive strategy of employing technological progressivist language, by highlighting the application of regulatory science—a strategy within the administrative pragmatism discursive category—as a prerequisite to making conclusive statements about nanotechnology's benefits.

The overlap statements also diffuse power more broadly across the nanotechnology risk discourse. Because the statements I have identified as overlapping typically "overlap" because speakers tend to turn to science to support their arguments, such statements diffuse through the nanotechnology risk discourse language suggesting science's appropriation of other, non-risk-related social concerns. Just as Bruno Latour's foray into the nineteenth-century French countryside demonstrated how Pasteur's laboratory shaped networks by co-opting society's interest in healthy cattle into Pasteur's own scientific interests, so too does the infusion of scientism into discourses co-opt broad social interests into the scientific enterprise. As Latour argues, displacing societal interests by moving them into the laboratory leads to scientific and non-scientific actants and actors sharing the same space, erasing the notion of being either inside or outside of science. The laboratory—or in the nanotechnology case, regulatory science—is a lever with which science pries society's interests out of their traditionally non-scientific grounding and exposes them to scientific inquiry, inscribing them to the body of scientific knowledge.[43] This displacement, to use Latour's term, drives the entire nanotechnology environmental policy debate toward a narrowing of terms that fit within the narrative of scientism and regulatory science.

Blurred Concepts, Discursive Strategies

The third study finding that overlap statements blur boundaries between discursive categories is supported by observations such as Potter and Wetherell's that people will use language in complex and creative ways to achieve objectives within discourse, and Hajer's articulation of how a broad narrative can travel through discourses and shape behaviors within discourse.[44] Categories of thought or message—discursive categories—therefore are shaped by story lines, scientism in the nanotechnology case, and thus shaped are employed within discourse to achieve political objectives. In the policy context this indeed is interesting and worthy of STS exploration. However, what also may be interesting are individual statements that seem contradictory, or at the expense of making a point to a particular audience seem to undermine the speaker's political strategy within the discourse. For example, while nanotechnologist Vicki Colvin's 2003 introduction of the term "wow-to-yuck factor" was meant to caution Congress about the potential for bad publicity to stifle nanotechnology development and commercialization, the term has now made its way into the literature to describe public discomfort with some new technologies.[45] The prominence of overlap statements in my analysis and the significant impact they have on messages conveyed in the nanotechnology risk debates suggest that this provides yet another opportunity for STS scholarship. As Dryzek has noted and as my analysis supports, exchanges of ideas can cross discursive strategies.[46] Yet overlap statements suggest that more is going on than just idea exchange. When voices that speak most fluently in the language of, say, technological progressivism, also make statements that incorporate administrative pragmatism language (or, more rarely but still observed in my analysis, risk society), such statements may warrant deep analysis to investigate their implications for power and boundary struggles.[47] Language of course is a highly personal attribute, and therefore overlap statements may be idiosyncratic to the individual speaker. This points to the need for deeper analysis of individual speakers, perhaps by combining discourse analysis with ethnography. A possible limitation of my analysis is its focus on nanotechnology risk statements. The deeper investigation of individual participants within discourse that I am suggesting ought to look more broadly at the individual's use of language. Doing so may bring out cultural implications not uncovered in a study as narrowly focused as mine.

Overlapping Language and Blurred Boundaries Offer Opportunities for New Thinking

This section analyzes how the thinking that is manifest in overlap statements may impact environmental policy. Specifically, I explain the implications of what the thinking of knowledge brokers, as such thinking relates to overlap statements, might suggest for the possibilities for widening the range of voices heard in nanotechnology risk debates.

The two overlap statements by Andrew Maynard and Bart Gordon examined near the end of chapter 3 showed, respectively, scientism in the form of regulatory science softening a risk society statement, and administrative pragmatism blunting a technological progressivist position. These two statements, along with a number of others analyzed in this study, illustrate how knowledge brokers' thinking about nanotechnology risks remains unsettled. Perhaps because the economic benefits of nanotechnology are still somewhat speculative, and on the other side of the benefit-risk equation, because as of this writing there has yet to be a single human illness or death that can be attributed to manufactured nanoparticles, positions have not yet become entrenched. People are still grappling to understand the implications, positive and negative, of nanotechnology.

This unsettled state may provide opportunities to engage new voices into policy discourse. People are more inclined to be receptive to different perspectives when their own views remain inchoate and new information will not put them at a social or material disadvantage.[48] So far, we appear to be at that place with nanotechnology. In chapter 6 I will suggest specific approaches for exploiting this apparent unsettled state of thinking by knowledge brokers. I will also argue that time is of the essence in putting such approaches into practice, since history suggests that such unsettled thinking in public policy is a temporary condition.

FINDING 4. INFLUENTIAL DISCOURSE PARTICIPANTS HAVE A COMMON SET OF SKILLS AND ATTRIBUTES

The nine individuals whose language was analyzed for this study have three important skills: expertise in the scientific subject matter, presence at key policy venues, and the capability to clearly translate risk issues into policy-relevant language. Analyses of other environmental debates show that *knowledge brokers*, to use Karen Litfin's term for individuals who can speak the language of science in terms understandable for policy making, play important roles in risk debates.[49] The construction of discursive categories has proven useful for evaluating how these individuals shift strategies within discourse to tailor messages to particular audiences, while maintaining their credibility as authorities on nanotechnology and issues related to environmental risks.

As I noted in chapter 3, in *The Scientific Life*, Steven Shapin argues that influential scientists—whom he labels *truth-speakers*—possess three characteristics that are central to their success in communicating knowledge: personal virtue, familiarity, and charisma.[50] In uncertain and unstable times, people come to trust not some objective set of "facts," which seem to be put into question with every new interpretation of each new piece of information, but rather they trust the personal attributes of individuals they admire.[51] Shapin's perspective is consistent with my observations of the individuals whose statements are the subject of this study. The three skills and attributes that make up this fourth study finding—expert, present, and articulate— are quite similar to Shapin's. My identification of these skills and attributes comes partly from my analysis of the nine individuals' statements and writings, but equally if not more from my personal interactions with them over the past five years within the nanotechnology risk discourse. The following discussion builds on the description of this study finding earlier in this chapter, focusing here on those considerations that are of particular relevance to the ability to advance policy initiatives.

Expertise in the scientific subject matter is not a particularly well-defined attribute. The nine individuals whose statements are highlighted in this book come to the nanotechnology discourse with different academic and professional backgrounds. Some can create nanoparticles, others have no laboratory training but have in-depth knowledge of how risk information fits within legal and/or commercial frameworks. However, what they all have in common is a general understanding of regulatory science. Hazard, dose-response, and exposure are familiar concepts not only to the scientists, but also to the lawyers and industrialists. And all can locate and articulate regulatory science concepts from their particular disciplinary and professional perspectives.

Being present at key policy venues means making oneself available for the numerous congressional hearings (both to testify and to observe), conferences, workshops, and other meetings that are held not just in Washington, DC, but throughout the country (often at the universities that receive nanotechnology grants through the NNI) and around the world. This entails significant requirements of time, resources and perhaps most important, energy. The inability to meet one or more of these requirements likely winnows out many potential participants in the nanotechnology risk debates.

Finally, the capability to clearly translate risk issues into policy-relevant language is a valuable and in my view, uncommon, skill. Such translation within the congressional environment is a good example. Given the many and diverse demands placed on members of Congress and their respective staffs, it is difficult to find a member or staff person who understands the language of risk and regulatory science, except possibly within the House Committee on Science and Technology. In the 665 statements used in my analysis, including some presented in chapter 3, there are examples of how individuals such as Vicki Colvin and Andrew Maynard are able to take complex regulatory science concepts and present them in ways that are understandable and relevant to a particular hearing topic or member's interests. Of course, the policy arena is larger than Congress or even government, and therefore the translation skill is equally important for communicating in blogs, press interviews, and conference presentations. Regarding conferences, what I would call "reverse translation" is also something at which these nine individuals are adept: they have the ability to communicate the interests of policy makers back to scientists, lawyers, and business people in language that each of these groups can understand. In fact, all nine of these individuals can do such reverse translation for any one of these three groups, as well as when members of these different groups are gathered together at one venue.

This finding concerning influential discourse participants presents opportunities for further exploration in knowledge broker or policy entrepreneur scholarship. However, the importance of recognizing these three characteristics lies mostly in understanding how such abilities can be applied to advance policy discourse and for the nanotechnology case, how debates can be broadened and enriched. As such, the finding that influential discourse participants have a common set of skills and attributes is useful in chapter 6's proposal for targeted policy initiatives.

Power Relations

The power implication of this finding is that engagement in policy debates may be difficult for those who do not have expertise in the scientific subject matter, cannot be present at key policy venues, or have not developed the capability to clearly translate risk issues into policy-relevant language. It may be, as described by Litfin, that people with these skills—knowledge brokers who understand and are able to articulate how science fits within political and economic constructs, and so are able to align their knowledge with the interests of the actors who create and shape policy—are in an advantageous position to advance political strategies within discourse.[52] Each of the nine individuals whose statements were used in the discourse analysis is steeped in either the science of technology development, regulatory science, or the use of science and technology in legal and/or business contexts. Their skills are supported in the nanotechnology case by a regulatory science narrative within which they operate comfortably. This situation supports Litfin's contention that in addition to knowledge brokers' skills, their influence also depends on the context within which debate occurs.[53]

This finding's power implications raise both challenges and opportunities. If experience within the chemicals regime is any indication, it is unlikely that the public policy arena will open itself up to individuals who do not possess the same or similar skills and attributes I have identified in this study.[54] That said, knowledge brokers are made, not born. Interventions are possible to bring additional perspectives into risk debates by developing new knowledge brokers from communities and disciplines not typically represented in risk debates. Strategies and actions that make inroads in at least partially decentering regulatory science also may create openings for new influential participants in risk discourse.

Skills, Attributes, and the Use of Symbols

There exist significant opportunities to advance STS scholarship in the area of influence on policy debates. To complement approaches such as actor-network theory, which focus on the settings in which science takes place and how science positions itself in society, and feminist studies, which by examining science as a ritualistic social institution seek to understand how societal views of gender have influenced the conduct and findings of science,[55] I suggest that conducting ethnographic research into the symbolism embraced and articulated by knowledge brokers could lend insights into these individuals' influence over policy debates.

As described by Spradley, *symbolic interactionism* is a theoretical construct to explain how the meanings individuals give to things (symbols) can help explain their behavior in social interactions. Within this construct, culture is a cognitive map that guides individuals as they interact with one another and interpret symbols.[56] The US environmental policy arena is a particular cultural setting, with symbols such as data, protocols, methods, tools, benefits, and of course, risks. People who are skillful at wielding power within discourse exercise their influence through interpersonal interactions, and through such interactions not only follow but shape the cognitive map used by participants in policy debates. Therefore, what could be useful to further STS investigation of environmental policy debates is an ethnographic analysis focused on risk-related symbolism—how symbols appear in the cognitive map of the policy arena, how influential people shape the meaning of these symbols, and how symbolism impacts the terms and outcomes of policy debates.

Engaging in the Policy Debate

First and foremost, for individuals to successfully engage in policy debates, they must be recognized by their environmental policy peers as knowing what they're talking about. At first blush this may seem obvious—expertise in general is highly valued in society. However, I am referring to expertise in a broader sense. Not only must the successful knowledge broker be expert in her or his professional field, but she or he also must be conversant in all aspects of policy making. For instance, scientists must understand how variation in nanoparticle properties relate to testing provisions in the Toxic Substances Control Act; attorneys should be well versed in both the legal and scientific implications of chemical test guideline amendments; it is taken for granted that government officials are current on the state of nanomaterial commercialization; likewise, industrialists are assumed to be following nanotechnology-related legislative and regulatory developments. The nanotechnology knowledge broker must be able to see the entire environmental policy chessboard.

Of course, seeing the game board is not sufficient: to be influential, one who aspires to participate in risk discourse as a knowledge broker also must be an active player in the policy arena. She or he must *be present* in the community. Nanotechnology knowledge brokers will attend the congressional hearings (whether or not they themselves are testifying); give presentations at major conferences (just attending is not sufficient); and they will write articles and blogs, get published in journals, and give press interviews. They will do all these things frequently and—most important for nanotechnology—they will do them globally. In an emerging technology field where developed and developing nations alike are looking to find economic opportunities, being present and relevant in the United States policy arena means being active and recognized within the international nanotechnology governance community. It is not uncommon for active participants in the nanotechnology risk debates to encounter one another at a congressional hearing in Washington, DC, meet again a month later at a trade conference in Tokyo, and see each other a few months later at a meeting in Frankfurt.

It is my observation that people who are effective in the policy arena are effective explainers and story tellers. In congressional testimony, they frequently will describe scientific, business, or regulatory issues by referencing particular cases from a member's home state—for instance, citing a university-business collaboration in Oregon, or new funding for a next-generation lithium-ion battery plant in Michigan—to make their arguments relevant and personal. Rarely will they say that they cannot answer a particular question because "I'm just a chemist" or "that's outside my expertise as an attorney." Rather, they will tie their broad knowledge of environmental policy back to their professional expertise, giving a response such as, "As a scientist, I cannot, and I don't think it is my place, to judge the risk-benefit of any technology I develop. That is actually the policy makers' and the public's place. But it is up to me to provide hard data, and so that is what

I work towards. . . . You present to the public the benefits, they will make the right decision."[57] In this statement, Vicki Colvin asserts her expert position in generating "hard data," but also demonstrates an appreciation for the policy arena and the role of scientific information in liberal democratic society.

Consistent with Shapin, Litfin, and others who have studied the personal attributes of those who influence policy making, I have found that those who successfully engage in the nanotechnology environmental risk debates have important skills and attributes that others wishing to enter into, and be effective in, these debates also must have. To help identify a path forward for enhancing policy deliberations, this finding has pointed to the need to explore what it takes to bring new knowledge brokers into discourse as effective influencers of policy. My hope is that this exploration, as described here, will help illuminate this path forward and guide the actions I propose in chapter 6.

CONCLUDING THOUGHTS ON IMPLICATIONS

There are two overriding messages from this book's findings with respect to their implications for power relations, STS scholarship, and targeted policy initiatives. First, the domination of regulatory science is presently impacting how nanotechnology debates are being conducted—in particular, how risks are being framed and who is doing the framing. Its continued domination will present significant barriers to successfully applying STS scholarship and targeted policy initiatives to the goal of broadening and enriching risk debates. Scientism, through regulatory science, is the single most-powerful narrative running through the environmental risk epistemic community.[58] However, there is a second critical message that could work against the first: the nanotechnology risk debate is unsettled, even for knowledge brokers, and there are opportunities to enlarge and deepen policy discourse on nanotechnology and the environment. Major interest groups have not yet articulated firm positions on nanotechnology risks, and nanotechnology has yet to emerge as a strong engine of motive power for global capitalism.

Before proceeding to the next chapter's conclusions and recommendations, one additional set of questions bears considering. I have focused in this chapter on the implications of my study findings. What we do not know for certain, however, is *implications for what?* There are at least three questions one could pose regarding what, if anything, the nanotechnology case implies for environmental policy.

A first question might be whether the nanotechnology case is essentially a one-off event in environmental policy, unique in its issues and therefore neither an extension of past practice nor a harbinger of future risk issues.

I do not believe this is the case. The design and synthesis of manufactured nanomaterials has come about through a centuries-long evolution of technological capability to observe and manipulate matter. There is no reason to believe that this evolution will not continue and lead to other applications and products that will bring with them their own environmental concerns.

One also might ask whether nanotechnology is essentially an extension of the chemicals regime and as we learn more about nanoparticles, they will become just the latest set of ingredients in the chemical soup that society sups in the name of modernity. This outcome is more plausible than the first. However, any playing out of this scenario likely would be restricted to the use of nanomaterials in industrial applications or consumer products. I do not believe the use of nanoparticles in foods or food packaging, medicines, or medical devices will play out in the policy arena in the same manner that contaminants are addressed in the chemicals regime. Intentional ingestion or placement of substances into the human body is not an aspect of environmental discourse with which the chemicals regime has much experience—even with pesticides, human dietary exposure to chemical residues is considered an incidental and not an intentional outcome of chemical use. Therefore, under this scenario we could see a bifurcation of the nanotechnology risk discourse: one avenue of debate subsumed under chemical risk discussion, and a separate debate on intentional human exposure to nanoparticles.

A third possibility is that nanotechnology will be, or has become already, the start of something new in environmental policy discourse around emerging technologies. This is what I suggest in the first pages of this work, and this is the scenario I see as being most likely. The Venter Institute's creation of a self-replicating bacteria cell likely is the start of activity in synthetic biology that may rival or surpass the current enthusiasm for nanotechnology. From a human-needs perspective, these kinds of technological innovations likely will be driven by food and water shortages, new disease pressures arising from more-rapidly increasing ecological change and the greater mobility of human populations, and increased energy demand. From a human-wants perspective, just the single example of the consumer electronics market suggests that there will be increased demand for smaller, faster, and more entertaining applications and devices. In this area, DNA computing has been around for nearly a decade and is generating increasing interest.[59] If we are not leaving the chemicals regime behind entirely, it does appear that we are moving into technological fields that are creating and incorporating into products very small, complex, and in some cases living materials that may become ubiquitous in society. If this is so, then the nanotechnology risk discourse may be reflective of the start of new types of debate around the impacts of emerging and future technologies.

NOTES

1. Foucault, *History of Sexuality*, 100.

2. For Dryzek, what distinguishes green romanticism from green rationalism, both of which he locates within green radicalism, is that the former rejects the Enlightenment belief in science as the principal source of human knowledge, while the latter embraces the role of science in human development. Dryzek, 154.

3. HistoryLink.org, "WTO Meeting and Protests in Seattle (1999)—Part 1," (HistoryLink.org Essay 9183) and "WTO Chronicles: Nancy Pennington Talks About Turtles," (HistoryLink.org Essay 2871), http://www.historylink.org.

4. Meadows, Donella, Jorgen Randers, and Dennis Meadows, *The Limits to Growth*, New York: Universe Books, 1972. See NASA's use of the term "big blue marble" in its description of its environmental observation activities, "Blue Marble Next Generation," http://earthobservatory.nasa.gov/Features/BlueMarble.

5. Merchant, 59.

6. Wendy Brown, *Regulating Aversion: Tolerance in the Age of Identity and Empire* (Princeton: Princeton University Press, 2006), 15.

7. Latour and Woolgar, 150.

8. Eileen Crist and H. Bruce Rinker, "One Grand Organic Whole," in *Gaia in Turmoil: Climate Change, Biodepletion, and Earth Ethics in an Age of Crisis*, edited by Eileen Crist and H. Bruce Rinker (Cambridge: The MIT Press, 2010 [Kindle version]), 4.

9. Eileen Crist and H. Bruce Rinker, "One Grand Organic Whole," in *Gaia in Turmoil: Climate Change, Biodepletion, and Earth Ethics in an Age of Crisis*, edited by Eileen Crist and H. Bruce Rinker (Cambridge: The MIT Press, 2010 [Kindle version]), 7–8.

10. Eileen Crist and H. Bruce Rinker, "One Grand Organic Whole," in *Gaia in Turmoil: Climate Change, Biodepletion, and Earth Ethics in an Age of Crisis*, edited by Eileen Crist and H. Bruce Rinker (Cambridge: The MIT Press, 2010 [Kindle version]), 7–8.

11. Karen Litfin, "Principles of Gaian Governance: A Rough Sketch," in *Gaia in Turmoil: Climate Change, Biodepletion, and Earth Ethics in an Age of Crisis*, edited by Eileen Crist and H. Bruce Rinker (Cambridge: The MIT Press, 2010 [Kindle version]), 198.

12. Eileen Crist, "Intimations of Gaia," in *Gaia in Turmoil: Climate Change, Biodepletion, and Earth Ethics in an Age of Crisis*, edited by Eileen Crist and H. Bruce Rinker (Cambridge: The MIT Press, 2010 [Kindle version]), 323.

13. Litfin in Crist and Rinker, 199.

14. Foucault, *History of Sexuality*, 143.

15. The latest edition of the *Handbook of Science and Technology Studies* includes a chapter on "Anticipatory Governance of Nanotechnology." Edward J. Hackett, et al., editors, *Handbook of Science and Technology Studies, 3rd Edition* (Cambridge: The MIT Press, 2007), 979–1000. The Center for Nanotechnology in Society at the University of California, Santa Barbara, is one center of nanotechnology-related STS research (http://www.cns.ucsb.edu), as is the Center for Nanotechnology in Society at Arizona State University (http://cns.asu.edu).

16. Jasanoff, *The Fifth Branch*, 39.

17. Jasanoff, *The Fifth Branch*, 40.
18. Beck, *Risk Society*, 59.
19. Kleinman, *Science and Technology in Society*, 3.
20. Jasanoff, *The Fifth Branch*, 228.
21. Gieryn, 43.
22. Gieryn, 46.
23. By regulatory science "community," I mean all those who have a stake in the production and use of regulatory science information. Regulatory agencies, academic and government scientists, the legislative and judicial branches of government, industry, and public interest groups are members of this community.
24. National Research Council, Committee on Toxicity, *Toxicity Testing in the 21st Century: A Vision and a Strategy* (Washington, DC, The National Academies Press, 2007), 98–119.
25. Jasanoff, *The Fifth Branch*, 14.
26. Christopher J. Bosso, editor, *Governing Uncertainty: Environmental Regulation in the Age of Nanotechnology* (Washington, DC: RFF Press, 2010), 3–4.
27. There are more than 84,000 chemical substances on the TSCA Chemical Substances Inventory; for only a small fraction of those has EPA received sufficient data to make risk determinations in accord with EPA's own risk assessment guidelines. On average, about 700 new substances are added every year. Information on the TSCA inventory may be found at http://www.epa.gov/oppt/newchems/pubs/invntory.htm. Also see US Government Accountability Office, *Chemical Regulation: Options Exist to Improve EPA's Ability to Assess Health Risks and Manage Its Chemical Review Program*, GAO Report GAO-05-458, June 2005, http://www.gao.gov/new.items/d05458.pdf.
28. Natasha Gilbert, "Crucial Data on REACH Not Disclosed," *Nature News*, Nature.com, 21 April 2010, http://www.nature.com/news/2010/100421/full/4641116a.html.
29. Bosso, *Governing Uncertainty*, 136.
30. EPA's 1989 attempt to ban asbestos from products was overturned in 1991 by the Fifth Circuit Court of Appeals because, in essence, the court determined that EPA had not provided a sufficient regulatory science justification for the ban. See http://www.epa.gov/asbestos/pubs/ban.html. For a concise summary of the issue, see Environmental Working Group, "The Failed EPA Asbestos Ban," http://www.ewg.org/sites/asbestos/facts/fact5.php.
31. Marc K. Landy, Mark J. Roberts, and Stephen R. Thomas, *The Environmental Protection Agency: Asking the Wrong Questions* (NY: Oxford University Press, 1990), 200.
32. Sunstein, 35–37.
33. Leo Marx, *The Machine in the Garden: Technology and the Pastoral Ideal In America* (New York: Oxford University Press, 1964), 43.
34. Sheila Jasanoff, *Designs on Nature: Science and Democracy in Europe and the United States* (Princeton: Princeton University Press, 2005), 133.
35. Foucault, *History of Sexuality*, 100–101.
36. Mills, 16.
37. Sunstein, *Risk and Reason*, 110–111. Sunstein uses the term "cost-benefit analysis" broadly to include risks as well as the distribution of benefits across society.

38. For information on the court case and endangerment finding, see US EPA, "Endangerment and Cause or Contribute Findings for Greenhouse Gases under Section 202(a) of the Clean Air Act," http://www.epa.gov/climatechange/endangerment.html.

39. US EPA, "November 3–6, 2009: Assessment of Hazard and Exposure Associated with Nanosilver and other Nanometal Pesticide Products: A Consultation," http://www.epa.gov/scipoly/sap/meetings/2009/110309ameeting.html.

40. Andre E. Nel, et al., "Understanding biophysicochemical interactions at the nano–bio interface," *Nature Materials* 8, 543–557 (2009), published online: 14 June 2009, http://www.nature.com/nmat/journal/v8/n7/abs/nmat2442.html.

41. Philip Ball, "Synthetic Biology for Nanotechnology, *Nanotechnology* 16, R1-R8 (2005). Published online: 8 December 2004, http://iopscience.iop.org/0957-4484/16/1/R01.

42. Haack, 23. Philip Kitcher, *Science, Truth, and Democracy* (Oxford: Oxford University Press, 2001), 175.

43. Bruno Latour, "Give me a laboratory and I will raise the world," in *The Science Studies Reader*, M. Biagioli, editor (258–275), 266.

44. Potter and Wetherell, 47; Hajer, 56.

45. Gary Marchant, Ann Meyer, and Megan Scanlon, "Integrating Social and Ethical Concerns Into Regulatory Decision-Making for Emerging Technologies," *Minnesota Journal of Law, Science, and Technology*, 2010:11(1): 345–363, 346.

46. Dryzek, 8

47. Of the thirty overlaps in the 269 technological progressivism statements, twenty-four were overlaps with administrative pragmatism; only six were with risk society, and those six came from the two scientists.

48. For an extensive discussion of one scholar's perspectives on the formation of public opinion and positions, see Cass Sunstein, *Going to Extremes: How Minds Unite and Divide* (Oxford: Oxford University Press, 2009).

49. Litfin, 37. Also see Bosso, *Pesticides and Politics: The Life Cycle of a Public Issue*.

50. Shapin, 5.

51. A complementary perspective is Yaron Ezrahi's concern with *outformations*. Because outformations are not created with any clear methodologies, they are unavailable for public analysis. He suggests that low-cost (in terms of effort required for their interpretation) outformations indicate a postmodern shift from reasoned discourse toward speculative decision making (Ezrahi in Jasanoff, *States of Knowledge*, 254, 260).

52. Litfin, 188.

53. Litfin, 188.

54. A review of past and present membership of key science advisory panels, such as the National Academy of Science Board on Environmental Studies and Toxicology (http://dels.nas.edu/best), or the EPA's Science Advisory Panel (www.epa.gov/sab) suggests that the skill sets and areas of disciplinary expertise of those chosen to provide advice on regulatory science within the chemicals regime is relatively narrow and quite well defined.

55. Sandra Harding, *Whose Science? Whose Knowledge?* (Ithaca: Cornell University Press, 1991), 87.

56. James P. Spradley, *The Ethnographic Interview* (New York: Holt, Rinehart, and Winston, 1979), 6–7.

57. US Congress, House Science, 9 April 2003, 81.

58. Halfon, 19. Citing Peter Haas (1992), Halfon defines *epistemic communities* as "loosely linked but powerfully situated international communities of experts and policy-makers who share a common technical understanding of a problem."

59. D. I. Lewin, "DNA Computing," *Computing in Science & Engineering,* May/June 2002, vol. 4, issue 3, 5–8.

6

Conclusions and Proposed Path Forward

This final chapter serves three functions. First, it provides my conclusions regarding discourse analysis as an analytical approach and the nanotechnology case as a vehicle for applying that approach. Here I build on chapter 5's discussion of the nanotechnology case's implications for power relations, STS scholarship, and targeted policy initiatives, to provide specific conclusions about what the nanotechnology risk discourse means for these three areas. Second, it presents a proposal for broadening and enriching environmental policy discussions about the impacts of emerging technologies. Third, the chapter concludes with final thoughts on the future directions of discourse on risk and emerging technologies.

CONCLUSIONS ON THE USE OF DISCOURSE ANALYSIS IN THE NANOTECHNOLOGY CASE

I have approached the nanotechnology case from the perspective, as articulated by Margaret Wetherell, that language is not passive but rather is socially constructive.[1] That is, language plays an active role in building our social reality. This perspective, joined with my adoption of Foucault's view that knowledge and power are produced in discourse through language and practice, has informed my belief that discourse analysis is a useful tool for examining nanotechnology risk debates.[2] This belief has been justified through this study. The exploration of risk discourse in the nanotechnology case has provided useful insights into the operation of power relations within risk debates, as well as how STS scholarship and targeted policy initiatives might be employed to improve democratic participation in environmental policy making. These insights have led to several conclusions. I begin with conclusions on power relations, and give power the most thorough treatment because its implications are the most generalizable beyond risk debates to the broader realm of public policy. Next I discuss conclusions for STS scholarship, indicating where I think opportunities exist to enhance the use of discourse analysis as a social science tool. Finally, I present what I believe is the principal public policy conclusion derived from my analysis: that environmental discourse ought to put less emphasis on risk and more on enriching the human and environmental condition. This conclusion serves as the point of departure for the policy proposal in the chapter's second section.

Power Relations

This book has given significant treatment to the relationship between language and power, and how this relationship is affected by the generation and use of scientific knowledge to understand environmental risks. While the nanotechnology case has served as the vehicle for this study, the implications go more broadly to how public policy is conducted in the United States at a minimum, and likely in most if not all liberal democratic societies. Language matters. Not only what is said but who is allowed to say it has an impact on the health and quality of, and trust in, democratic government. A measure of the state of a democracy is the extent to which power over life and death—biopolitical power—is distributed among the citizenry. The previous chapters have helped develop an argument that currently, at least in terms of how public policy addresses the introduction of new technologies into society, biopolitical power is not optimally distributed, but that the nanotechnology case points to a few promising paths forward to richer discourse, better policy development, and healthier democracy.

Presently, power is concentrated in those groups or individuals who are adept at using the language of regulatory science, in particular the use of that language to articulate arguments around quantitative assessments of risk in accord with regulatory science's framework of technical risk assessment guidelines and procedures. It is not just that technical expertise is required to produce or describe regulatory science's construction of environmental risk. Rather, the entire construction of risk defines what knowledge and perspectives are allowed into policy debate. For instance, if *hazard* is described solely in terms of the probability of disease, then a number of other possible negative impacts are omitted from discussion. Where within such dialogue is there a place for other impacts, such as the cultural disruptions that new technologies may cause or the concern that the technology will make an already complex modern society even more complicated?[3] No such discourse occurred, in a serious way, with the introduction into society of the personal computer or the cell phone. The nanotechnology case suggests that the American public policy arena may be prepared to engage in such discourse; if so, there are important implications for biopolitical power.

Discourse analysis has been highly useful in helping draw from the nanotechnology case two important conclusions regarding biopolitical power. First, the power relations identified in my analysis support Foucault's articulation of power's *productive* aspect—that power is not only repressive, but also can be useful and beneficial.[4] As noted in earlier chapters, the dominance of scientism in the environmental protection episteme, and regulatory science in risk discourse, suggests a concentration in government of the power to define, identify, and address environmental risks from nanotechnology. I believe this can be positive for American democracy, assuming (as I do) that in liberal democracies accountability for wielding biopolitical power is greater with institutions of government than it is with institutions that are accountable not to the citizenry but to particular private interests. At issue is whether government institutions are capable of wielding this power constructively for social benefit; I will come back to this in my proposal for enhancing policy debates.

A second conclusion related to biopower revealed through the nanotechnology case is that there is what Foucault called the "micro physics" of power that can be drawn upon to enhance risk debates.[5] This micro physics relates to the small, repeated behaviors and practices that people engage in, at all levels throughout society, that provide the force behind what Foucault characterized as the capillary movement of power, circumscribed by and defined within discourse.[6] Examples of such capillary movement illustrated in this study's nanotechnology risk statements include experts giving testimony before Congress, government agencies hiring communications experts to develop and disseminate nanotechnology-specific messages, NGOs setting up nanotechnology blog spaces, and companies creating alternative forms of specific nanoparticles and testing them in cell cultures. Using the study of language to locate these individual, repeated behaviors within discourse, and being able to understand these practices in terms of their influence on both language and power relations, has helped reveal specific activities that may have the potential to either limit or broaden debate on environmental risks.[7] Rooting power in practice and knowledge creation, and recognizing that individual activities—repeated in ways consistent with the discourse within which they are located—make a difference in the nature of power relations, may lend optimism to the belief that altering specific behaviors can effect change in certain knowledge-creation practices that may not presently be optimal for conducting rich, broadly informed policy debates.

Some may argue that policy debates on environmental issues are circumscribed by the statutory frameworks within which governments regulate environmental protection. This argument puts the legal cart before the horses of governance. Environmental governance is intended to protect citizens and other living entities that do not have a voice in human society—possibilities for protection ought not be limited by current interpretations of existing laws. Even within existing environmental statutes, there is room for discussion of risk that is richer than that provided by regulatory science. The Toxic Substances Control Act defines risk as "unreasonable risk of injury to health or the environment."[8] Other statutes are similarly broad. Our laws do not limit debate. Neither does science limit debate, when science is understood in terms of its Enlightenment roots as the pursuit of understanding. What limits discourse is our strict adherence to a calculated system of governance, reinforced by myriad micro behaviors practiced within the scientism narrative. Once we know what micro physics are at work within discourse, we can leverage them to make calculated governance work for, rather than against, our objectives for enriching and broadening risk debates.

I have discussed how regulatory science, as an expert system and technology of governmentality, reinforces our system of calculated governance. Science and law jointly support calculation, and in doing so not only shape liberal democracy but also support global capitalism.[9] Yet there is some push-back with nanotechnology. A version of Beck's risk society, albeit in the United States a fairly weak version, emerges from the analysis of nanotechnology risk discourse. This has revealed a breach in regulatory science, and possibly even of calculated governance.

This breach has power implications. Both industry and environmental NGOs have shaped their strategies for interacting with government policy makers around regulatory science's risk paradigm. For nanotechnology, we see a focus on discussions of whether or not toxicology studies provide indications of hazardous effects. For instance, detection and measurement of particles in the environment is a growing area of interest. Regulatory science therefore holds considerable power over debate. Letting in other concerns, and therefore other areas of expertise, would dilute that power.

The power implications are less clear for government or for intergovernmental and quasi-governmental bodies. There indeed are significant regulatory science interests within these institutions, particularly in executive branch agencies. However, these organizations generally have mandates that are broad enough to cover issues outside the scope of regulatory science. For them, the issue may be less of power than of expertise, structure, bureaucratic inertia, and—perhaps most important—defending the legal standing of their positions.

Global capitalism has no particular interest in the relative power position of risk within regulatory science, so long as regulatory science can provide certainty in the administration of legal frameworks that facilitate industrial development and commerce. However, as a social force as well as a network through which power flows, capitalism shapes how power relations are manifest. Capitalism does not require that benefits (and risks) be solely articulated in quantitative terms. However, it does demand consideration of production, markets, and profit. Weber's iron cage/hard shell is one metaphor for capitalism; one that has strong cultural implications. From the perspective of discourse, one also could view global capitalism as casting out through society a network of strong behavioral and institutional threads that bind together a wide range of power considerations, including biopolitical power. This certainly seems to be the case in the nanotechnology debates, where few if any of the statements used in the case study challenged the desirability of pursuing the economic benefits of nanotechnology. It would seem to be difficult, if not impossible, to consider power over life and death without taking into account the network of global relationships that define and drive the interdependent economic activities upon which so much of humanity now depends for its livelihood.

STS Scholarship

Chapter 5 describes several implications of this book's findings for STS scholarship, and suggests a number of lines of STS inquiry that could be pursued to better understand those implications. It is clear that I believe applying discourse analysis to a case such as nanotechnology can open up issues for pursuit through STS scholarship. In this chapter, I wish to describe two conclusions for STS scholarship that I have arrived at through the nanotechnology case.

My first conclusion is that further, and perhaps deeper, studies of the language of public policy can advance both STS as an interdisciplinary endeavor and discourse analysis as a tool for social scientists. My analysis of nanotechnology risk language is useful in drawing out a dominant narrative, such as regulatory science, and providing a means for identifying important relationships, such as that between natural entities and human-created artifacts, and describing the implications for those relationships on power relations and other aspects of policy debates. In general, these are "what" considerations; they focus less on *how* relationships form, power is exercised, and narratives come to be dominant within the environmental protection episteme. Other types of STS analysis may help reveal more of the "how" considerations in public policy.

By a deeper study of language, I mean in particular a closer contextual examination of how different individuals articulate central concepts in areas of governance such as environmental policy. For instance, Litfin asserts that "the discursive worlds of experts and policy makers are inherently different."[10] Yet my analysis of nanotechnology risk language suggests that such an assertion may warrant further exploration. The universal acceptance of regulatory science across the nanotechnology risk discourse implies that policy makers and experts may at times occupy the same discursive space, or at the very least their spaces overlap. If this is true, then the inherency part of her assertion also could be examined. The point is not to challenge Litfin's specific position, but rather to note that further unpacking might be done of relationships between actors. Exploring the language used in interactions between scientific experts and policy makers, for example, could provide insights not just into what their respective positions are regarding the generation and use of knowledge to inform decision making, but also into *how* each shapes the other's views on knowledge creation and use.

My second conclusion is that there is much opportunity in STS to build off existing scholarship to explore how environmental discourse is characterized. In conducting the research for this book, it was useful for me to reshape other scholars' discourse descriptions to construct three discursive categories that fit within my understanding of the nanotechnology case specifically and more generally for environmental risk debates. However, as suggested by the large percentage of overlap statements, how I have defined the three discursive categories does not allow for straightforward binning of the statements into specific categories. It may be that no optimal boundary drawing can be achieved between language within discourse. If such boundary-drawing optimization is possible, I do not find within the discourse analysis literature ready tools for accomplishing it. Even if such tools were available or could be developed, it remains an open question as to whether such tools are needed, or even are desirable, for STS scholarship. The overlap statements do blur the boundaries between discursive categories, but this blurring is itself revealing, such as in what it reveals about the power of regulatory science within the nanotechnology risk discourse, and the strength of scientism as a story line running through the environmental protection episteme. Therefore, while I have found discourse analysis to be a useful tool, it merits further probing by scholars for its applicability to specific STS lines of inquiry, in particular regarding attempts to categorize language. Rather than off-the-shelf application of discourse analysis approaches found in the literature, STS scholars may consider exploring how such approaches may be either useful or problematic in determining whether boundaries need to be drawn around language (and practice) and if they do need drawing, how much boundary permeability or mobility will the analytical approaches accommodate.

Targeted Policy Initiatives

My principal conclusion for targeted environmental policy initiatives is that conscious steps should be taken to broaden emerging technologies debates beyond risk, focusing more on language about *enriching the human and environmental condition.* The strength of regulatory science within the American environmental protection episteme in my opinion cannot be taken on directly with much likelihood of success. As represented in the nanotechnology risk statements, interests across environmental governance institutions are heavily invested, and have strong vested interests, in the use of regulatory science data for decision making. The micro physics of power reinforce these interests through repeated language and practice centered around regulatory science. Calculations of risk are integral to regulatory science, and therefore neither do I see possibilities for removing risk estimates, qualitative or quantitative, from environmental decision making. However, expanding discourse in specific ways that introduce "enriching the human and environmental condition" into debates can dilute and de-center risk language, and therefore weaken the hold that risk has on the environmental protection episteme as it relates to emerging technologies.

There are two main reasons why I believe it is important to weaken risk-related language and practice within environmental discourse. The first reason is because all the resources put into estimating the risks of hazardous substances over the past several decades have done little to reduce controversies about environmental risk, or to strengthen public trust in governance institutions' ability to protect citizens from harm.[11] In fact, recent evidence suggests that current approaches may not be achieving appreciable gains in some important areas of human health or environmental protection.[12] The second reason concerns the context of nanotechnology and other emerging technologies within environmental policy making. Whereas the bulk of today's industrial chemicals were introduced unreflexively into society, without debate on either their benefits in comparison to other substances or technologies, or their potential to create adverse environmental consequences, environmental debates on emerging technologies are beginning to become reflexive.[13] However, for reflexive thinking to translate into actions that avoid environmental impacts before technologies are introduced into society, there will not be time to engage in lengthy disputes over risk information. The pace of technology development is too rapid, and risk-based processes are too slow.

My conclusion that it is important to shift language at least partially away from the language of risk, and toward discourse on enriching the human and environmental condition, is related to this movement toward reflexive thinking. Reflexivity, in my view, can be as much about exploring possibilities as about avoiding danger. Approaching discussions on the environmental implications of new technologies from this perspective opens up discourse to broad discussions of benefits as well as of risks. Such broad discussion includes distributional issues: who would be the recipients of a technology's benefits, and on whom would fall most of the impacts? It would also include comparing the new technology to other future or existing approaches, and asking whether we even ought to pursue developing the technology. Most important for the proposal I present in the next section, moving discourse toward enriching the human and environmental condition can help focus policy makers on specific, pragmatic actions. An example is real-time technology assessment.

Real-time technology assessment involves joining social science research with physical and natural sciences as technologies are being developed, to influence the embedding of social values in technology development.[14] To be fully reflexive, such an approach would consider whether the specific application of nanotechnology itself—for the particular envisioned use(s)—is likely to result in net societal benefits. A real-time technology evaluation approach for a new nanomaterial would evaluate not just environmental and human health impacts over the material's life cycle, including inputs (e.g., energy usage and starting materials) and emissions resulting from production of the material, but also social considerations to be factored into the design, use, and ultimate disposal or recycling of the products of the technology. For instance, would use of the new material advance the welfare of traditionally disadvantaged people, or would it widen the social welfare gap between them and wealthier societies? Fully implemented, real-time technology assessment would be inherently multi- and transdisciplinary, requiring integration of physical and social/behavioral sciences into forward-looking approaches for a sustainable societal relation with technology.

Risk would continue to play a role in reflexive practices such as real-time technology assessment. However, opening up discourse to broader perspectives and possibilities for societal benefit moves discussion away from binary, go/no-go decisions on technologies and the materials and products they introduce into society. Instead, the discussion moves toward possibilities within specific contexts—possibilities both for the specific technology or product under consideration, or for other means (both technological and non-technological) to address the particular problem or opportunity to which the technology or its alternatives might be applied. In essence, this new way of thinking moves problem formulation far upstream of technology development, asking early on what problem or opportunity for benefit is at issue, and what if any role is there for technology in solving the problem or achieving the benefit. The challenge is to move discourse in this direction through activities that are achievable, meaning that they will be understandable and acceptable to interests presently operating within environmental governance institutions. In the next section, I make a proposal that attempts to foster this shift in thinking about and articulating the relationship between emerging technologies and environmental protection.

PROPOSALS FOR ENHANCING ENVIRONMENTAL POLICY DEBATES

Here I present my proposal for broadening and enriching debates on environmental policy issues. I begin by describing what I believe constitutes a good public policy proposal. Next I highlight several proposals that already exist in the literature. Finally, I present my own proposal, explaining how it can play a role in moving environmental discourse in a positive direction.

Attributes of a Viable Proposal

Undertaking new public policy activities costs both time and money. People must be tasked to do new work, and typically funds need to be allocated to support the work. Therefore, those who are charged with making public policy decisions, particularly in US regulatory agencies where resources seem to be in a perpetual state of severe constraint, tend to ask hard questions of any proposal to initiate new activities. Using a policy analysis framework is one means of helping determine whether proposals are viable and make the best use of scarce resources. Many such frameworks exist. I have chosen the following set of questions, drawn from Stokey and Zeckhauser, as useful for evaluating the viability of proposed targeted policy initiatives:

1. What is the underlying problem the proposal seeks to address?
2. What alternative courses of action have been considered?
3. What are the consequences of pursuing each of the alternatives?
4. What are the measures of successful implementation of the chosen alternative?[15]

In this section I address questions 1 and 4, and touch on questions 2 and 3 in presenting my specific proposal.

The underlying problem that proposals must address is that environmental policy debates concerning emerging technologies are focused on a narrow view of risk, without inclusion of other perspectives on the environmental impact of technologies. If full, inclusive debates eventually led to framing the issues within a narrow risk construct, that would be one thing—at least one could feel that such an outcome would have been the result of transparent, democratic processes. However, it is quite another thing—and, I would argue, something detrimental to public trust in liberal democratic institutions—if such framing occurs in a closed, unreflexive setting. It is my belief that the current environmental policy arena too often is closed and unreflexive, which leads to my response to question 4, measures of success.

Successful proposals will lead to inclusion in environmental policy debates voices that broaden, beyond the current regulatory science perspective, consideration of what constitute environmental impacts from emerging technologies, and those voices will be reflected in policy making and governance decisions. Consistent with my problem definition, the success of those additional voices being "reflected" in decisions does not necessarily mean that decisions always will address specific concerns articulated by those voices. Decision makers will, however, have factored those concerns into the decision-making process—something I argue currently occurs with much too little frequency.

Existing Proposals

Many advocates for enhancing environmental policy discourse on emerging technologies speak in general terms about public engagement and stakeholder involvement, without providing either specific proposals on how to make such engagement happen or evidence that such engagement has impacted policy making. For instance, in its 2006 *White Paper on Nanotechnology Risk Governance*, the International Risk Governance Council (IRGC) devotes an entire section to stakeholder participation. However, while the report recommends such approaches as round tables, negotiated rulemakings, and citizen juries, it does not suggest specific mechanisms through which these approaches could be implemented so as to impact public policy.[16] This is not a criticism of the report, but rather to point to the need for proposals that transform the ideas from sources such as the IRGC into specific, actionable activities for policy makers and others involved in environmental governance. There have been numerous proposals that are specific in their approaches to broaden debate. They are too many to describe here. Instead, I have chosen three that illustrate what can be found in the literature. They exemplify three common characteristics of such proposals, one positive and two negative: (1) they offer promising approaches for engaging voices not currently heard in policy debates; (2) they provide no specific mechanisms for implementation on a scale, both temporal and spatial, that would influence national policy making; and (3) they offer little or no evidence that their approaches would have the desired impact on public policy.

Maria C. Powell and Mathilde Colin initiated a University of Wisconsin project that led to the creation of the Citizens Coalition on Nanotechnology, which focused around Nano Cafes and web sites that offer citizens opportunities for education and engagement on nanotechnology-related issues.[17] The authors piloted the concept with six Nano Cafes in the Madison, Wisconsin area, and structured citizen engagement around the Danish consensus conference approach of experts engaging the public based on the public's questions, rather than the experts presenting what they know to the public.[18] While the pilot effort appeared to motivate citizens to engage in nanotechnology issues, the authors admit that their approach provided for no means to engage policy makers.[19] A useful next step could be development of a follow-on activity to link the Nano Cafe activity to local or national policy activities.

Richard Sclove calls for what he labels participatory technology assessment (pTA), also modeled on the Danish consensus conferences. Sclove argues that bringing citizens into the assessment of new technologies can lead to policy decisions that not only are more accepted across society, but also more effective and efficient.[20] Sclove provides four examples of where participatory technology assessment has been applied in the United States, and these examples are highly useful in describing the types of mechanisms that may be employed in American society. However, his examples show no evidence of having impacted national policy. Even more problematic is Sclove's inability to demonstrate that the model on which he has based pTA, the Danish consensus conference, has been effective in influencing public policy in Denmark. Of the twenty-four topics to which the Danes have applied consensus conferences, Sclove identifies only one policy issue, food irradiation, has having been directly impacted by the conferences and a second, genetic screening, has having influenced legislation.[21]

J. Clarence "Terry" Davies believes that an entirely new US product oversight system, to be created by merging six federal science and regulatory agencies into one, is needed to "move people from the bystander category to the informed category" with respect to technology oversight.[22] While Davies is fairly specific on how functions would be realigned to create the new oversight agency, he only goes as far as to indicate that new public participation approaches will be needed if the new oversight system is to operate in a transparent manner and consider societal issues beyond regulatory science in evaluating the impacts of new technologies.[23]

Taking the good ideas in these and other approaches as a point of departure, in the next section I turn to my proposal for broadening and enriching environmental policy debates. This proposal draws from the insights gained through the nanotechnology case; specifically, it addresses the conclusions for environmental policy described earlier in this chapter. It also is informed by my own environmental policy experience.

Proposal for Broadening and Enriching Environmental Policy Debates

My proposal involves developing and implementing a specific set of debate-broadening and enriching activities within an existing set of nanotechnology governance relationships, with federal agencies playing the leadership role in initiating these activities. While these activities hopefully will broaden risk debates generally across environmental governance institutions, given the findings of this study I have one priority objective for this set of activities: *to enlarge the perspectives of existing knowledge brokers, and to create new knowledge brokers with discursive competence in this broader way of thinking of technology's environmental implications.* In the following discussion, I begin by describing existing nanotechnology governance relationships. I then outline what specific activities I recommend, including how progress on these activities can be measured against their common objectives of broadening debates beyond risk and focusing more on language about enriching the human and environmental condition. I conclude with an argument for federal agency leadership.

Governance Relationships

While both within the United States and internationally it would be an overstatement to say that there is a nanotechnology governance *structure,* there do exist relationships within which there occur communication, coordination, and collaboration on issues related to nanotechnology environmental, human health, and safety concerns. How these relationships may look to federal agencies is illustrated in Figure 2, which provides an EPA perspective on how its nanotechnology relationships serve as information conduits to inform the decisions that EPA makes regarding the environmental implications of nanotechnology.

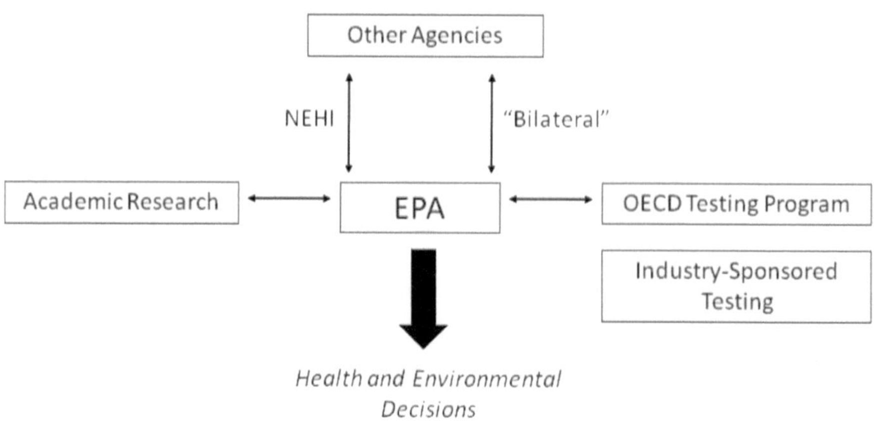

Figure 2. Nanotechnology Governance Relationships: An EPA-centric View

EPA's nanotechnology relationships, as illustrated in Figure 2, are largely described in chapter 1's more-general discussion of nanotechnology governance institutions. While this is an EPA example, it applies similarly to other regulatory agencies, such as CPSC, FDA, and OSHA. These federal agencies are active in the NEHI. Federal agencies also interact directly (described as "bilaterally" in the figure) with one another. For instance, the NSF and EPA work together to support academic centers conducting nanotechnology environmental research.[24] In general, each federal agency, whether it is part of NEHI or engaged more broadly in the NNI, has at least one person who serves as that agency's primary point of contact for nanotechnology-related issues. The OECD WPMN testing program, also described in chapter 1, provides federal agencies with a venue for collaboration with not only OECD member nations, but with other invited nations (Brazil, China, Russian Federation, Singapore, South Africa, and Thailand), industry through the Business and Industry Advisory Committee, and environmental and labor NGOs. Industry also interacts directly with federal agencies, either through regulatory programs or with federal researchers through cooperative research and development agreements established under the Federal Technology Transfer Act. The federal government directly funds academic research through its grants program, as well as interacts with academic researchers through collaborations between scientists. All of these relationships contribute to the development of regulatory science information that federal regulators use to inform their nanotechnology-related decision making.

Proposed Activities

Listed below are three highly targeted policy initiatives that can help move emerging technology discourse, as it relates to environmental protection, beyond risk toward greater emphasis on the broader theme of enriching the human and environmental condition. These activities are listed in a numerical order that indicates both sequencing and priority, both of which are important if resource-constrained federal agencies are to lead their development.

1. Hold one, or possibly a series, of regulatory science-focused meetings that bring STS scholars into the discussion of how to best inform a regulatory path forward for one or more specific nanomaterial applications. Time the meeting(s) to inform either the development or promulgation of specific nanotechnology-related regulatory actions. Agencies can leverage their relationships with the academic community to support the design and convening of the meeting(s) outside the federal government, to enhance the credibility of the meeting(s) as an independent scientific activity that nonetheless can, through federal agency leadership, impact regulatory

decision making. Success will be measured in the first instance by the degree to which the meeting proceedings reflect broader (i.e., beyond risk) considerations of impact, and ultimately to the extent those broader considerations are incorporated into the regulatory determination.

One could argue that just inviting STS scholars to meetings is not adding a substantive or meaningful enhancement of that community's engagement in environmental policy discourse, since most federal meetings are public and do not exclude STS scholars' participation. This is true. However, rarely are STS scholars invited to meetings concerning regulatory science with the express purpose of including their perspectives in the meeting agenda, discussions, and proceedings or reports. My proposal would entail doing just that. And in my experience, when a meeting agenda is intentionally constructed to draw into the discussion a particular discipline or area of expertise, those perspectives tend to end up being reflected in both the meeting discussion and any proceedings or report derived from the meeting.

2. Pilot real-time technology assessment within federal research organizations. These pilots would have two components: (1) academic grant solicitations that focus on research to advance real-time technology assessment approaches, and (2) an in-house element that brings social scientists into federal laboratories to partner with federal researchers who are conducting nanotechnology and other emerging technology-related regulatory science. A key relationship exploited in this activity would be one not shown in Figure 2 because it exists within the EPA box: the relationship between the research activities and regulatory programs within a single agency. In most cases, federal researchers interact regularly with their colleagues in regulatory programs, and this interaction can increase the potential for real-time technology assessment to have a direct impact on federal risk assessments and decision making. How much of this type of impact real-time technology assessment makes would largely determine the pilot's success. In addition, if the pilot were to use STS federal post-doctoral scientists as the social science participants in the real-time technology assessment pilot, success also could be measured by whether those post-docs, once they left their federal post-doc appointment, became active participants in the environmental policy arena.

A critique of this proposal could be that, in fostering academic-government collaborations, it neglects the true technology creators—the private sector. My response would be that the starting point and leadership for real-time assessment of technologies for their environmental and other societal impacts should be with the federal government. While the private sector has critical roles to play in the responsible development and deployment of technologies, only government has the mandate from, and responsibility to, society for evaluating and addressing the broad impacts of technology on public well-being.

3. Use the federal government's participation in the OECD to create a project that locates OECD chemicals-related activities within a larger "enriching the human and environmental condition" context. The WPMN has already initiated an activity, "Environmentally Sustainable Use of Nanotechnology," the purpose of which is to enhance knowledge about life-cycle aspects of manufactured nanomaterials, as well as positive and negative impacts on the environment and public health of certain nano-enabled applications at their different stages of development.[25] This new OECD activity would generate a set of case studies that examine specific nanotechnology applications from a societal "net benefit" perspective. This proposed activity would bring the WPMN testing program information (which essentially is regulatory science data) into these cases, after their initial development, to locate risk information within the larger context of net societal benefit. The intent is to de-center risk information, while acknowledging its continued relevance to decision making. Important to this project would be a series of workshops, perhaps using an existing relationship that the OECD has with the United Nations, to take this project's activities outside the OECD to developing nations. Doing so would both enrich the discussion as well as help develop a new cadre of participants in international environmental risk discourse. Success could be measured by how much international governance institutions reference the OECD project and whether such references find their way into policy debates in both national and supranational policy-making bodies.

A critique of the OECD is that it is a club of wealthy nations that does not always consider the concerns and needs of the rest of the world. Based on that view, one might consider the United Nations Environment Programme (UNEP) as a more-appropriate international venue for advancing nanotechnology policy discourse. Indeed, as noted above, UNEP has an important role to play. However, I also have noted that to date nanotechnology discourse resides within the chemicals regime. Because this is a targeted policy initiative—one intended to achieve tangible near-term results—I think it is wise to focus the effort where the energy around policy development is strongest. Internationally, the OECD is the international hub of government—and more broadly, governance—activity within the chemicals regime.

Why Federal Agencies

There are some who likely would challenge the notion that US federal agencies are best positioned to broaden environmental discourse. Some scholars, such as Habermas, have argued that the NGOs are well suited for this role.[26] However, I believe the nanotechnology case supports my view that in the United States not only are the NGOs, with some exceptions, not present in the types of governance activities that can be leveraged through proposals such as mine to broaden discourse, but they are as deeply entrenched in scientism and regulatory science as other key stakeholders.

It is my view that only the federal agencies have the ability to lead in fostering enhanced discourse on nanotechnology. Federal occupational, health, and environmental agencies have both the mandate of, and substantial experience in, engaging the public on the impact of environmental contaminants on health and well-being, and making decisions based on public input. While many if not all federal agencies may not be as well prepared as they could be to meet this challenge for nanotechnology, they are the best prepared of other alternatives.

These three activities cover a narrow range of possible actions and are intentionally very limited in scope. While their highly targeted nature suggests that their immediate impact will be circumscribed, the advantage of their narrow scope is that they directly leverage the micro physics of power and feed directly into the capillary network articulated by Foucault. They also depend on the leadership of US federal agencies that in the best of times are stretched for resources and operate in a highly charged political environment. However, it is my hope that not only will these particular activities be realized, but that they will be useful to other public policy scholars as examples of proposals that are both underpinned by theory and linked to existing institutional capabilities and decision-making mechanisms. There is room in public policy scholarship for targeted policy initiatives as well as broad thinking on advancing liberal democratic institutions and governance.

CLOSING REMARKS: RISK AND EMERGING TECHNOLOGIES

The nanotechnology case has proven useful in providing insights into how language and practice operate through discourse to shape environmental policy debates around the risks of emerging technologies, and how constructing such a case opens up possibilities for understanding power relations and other considerations in the function of government, governance and specifically regarding biopolitical power, governmentality. However, I want to emphasize in closing this work that nanotechnology is only a case, and in its current state of development may not be one of the most-pressing emerging technology issues that currently face environmental policy makers and society at large, or will face them in the coming years.

In the world of the very small, we may be entering into what David Rejeski calls "the molecular economy," where the manipulation of living and nonliving matter at the atomic and molecular levels will move beyond nanotechnology's current materials focus and into areas such as synthetic biology, molecular computing, and self-assembly of materials at the nanoscale.[27] The movement of technology in this direction raises ethical, legal, and other societal issues that make those identified so far for nanotechnology seem straightforward by comparison.[28]

We also face technological challenges in the world of the very large. If the dominant scientific views on the impacts of global climate change prove correct, there likely will be many ideas for large-scale technological manipulations of the environment to mitigate such impacts. There already are proposals to disperse clouds of nanoscale mirrors into the atmosphere to reflect solar radiation; others have proposed hoisting very large mirrors into space to accomplish the same objective.[29] Continued population growth will spur the design of technological fixes to problems of water availability, food production, and land use. These technologies will be considered within societal contexts that very likely will contain cultural stress, economic fear and opportunity (and opportunism), and political instability.

It is therefore all the more important that liberal democratic societies use cases such as nanotechnology as learning experiences to develop greater competence in establishing a sustainable and healthy relation with technology. We are learning that risk, language, and power are produced and move within discourse in a manner that, to date, has been unsuccessful in establishing such a relation. It may be time to speak less of avoiding or managing risk, since the language of risk has proven ineffective in focusing power's productive capabilities on advancing environmental protection and human well-being. Perhaps instead society is ready to direct the power of human ingenuity toward new thinking about technology's place in the better world we envisage for humanity and the environment. Giving deep thought to the language we use, and how we put that language to practice, may be a good place to start.

NOTES

1. Margaret Wetherell, Stephanie Taylor, and Simeon J. Yates, *Discourse Theory and Practice,* 16.

2. Stuart Hall, "Foucault: Power, Knowledge and Discourse," in Wetherell, Taylor, and Yates, 72.

3. As I write this paragraph, I am in an airplane flying through a turbulent air system. The plane is shaking violently enough to make typing difficult. This turbulence causes me, and I am sure other passengers, some anxiety. This anxiety does not negate the benefits I place on air travel, and (assuming we came through the turbulence intact and you are reading this) I have made many plane trips since this incident. However, using the technology of air travel creates a set of stresses nonexistent prior to its creation or not experienced by those who do not travel by air. These types of impacts are rarely, if ever, raised in discussions of the societal implications of specific technologies.

4. Hall, 77.

5. Hall, 77.

6. Michel Foucault, *Discipline and Punish* (London: Tavistock, 1977), 27.

7. In the nanotechnology case, the influence between language and power is bidirectional: the language of risk supports the power of regulatory science within the chemicals regime, and the power infrastructure that supports the chemicals regime sets the limits on the acceptable boundaries of risk language and its use in debate. My thanks to Jennifer Kuzma for pointing this out to me.

8. US Code of Federal Regulations, [15USC2603], [Page 1358-1362], Title 15, Commerce and Trade, chapter 53, Toxic Substances and Control, subchapter I, Control of Toxic Substances, section 2603, Testing of chemical substances and mixtures, (a)(1)(A)(i), [1977],USGPO Online: http://frwebgate.access.gpo.gov.

9. In the case of regulatory science, the supporting relationship between science and law is not only joint but *symbiotic* with respect to calculated governance. Regulatory science has come to justify its existence by citing the body of case law that has built up around safety levels set through quantitative risk assessment, and the legal system—which includes attorneys for regulators as well as regulated entities as well as the courts—has set quantitative risk assessment as the gold standard for justifying regulatory decision making. For instance, see the reference in chapter 5 to the asbestos case.

10. Litfin, 31.

11. Numerous studies have shown that the public's confidence in regulatory agencies has diminished in the past decade. For an example, see Hart Research Associates, *Nanotechnology, Synthetic Biology, & Public Opinion: A Report Of Findings Based On A National Survey Among Adults, Conducted On Behalf Of: Project On Emerging Nanotechnologies The Woodrow Wilson International Center For Scholars,* Washington, DC: Hart Research Associates, 22 September 2009, 14. The report may be found at http://www.nanotechproject.org/process/assets/files/8286/nano_synbio.pdf.

12. US Department of Health and Human Services, National Institutes of Health, National Cancer Institute, *Reducing Environmental Cancer Risk: What We Can Do Now,* President's Cancer Panel 2008–2009 Annual Report (Bethesda, MD: National Cancer Institute, April 2010), 97.

13. Examples of recent policy-focused documents with reflexive discussions of environmental impacts include the US EPA *Nanotechnology White Paper* and the numerous publications issued by the Woodrow Wilson International Center for Scholar's Project on Emerging Nanotechnologies (http://www.nanotechproject.org).

14. David H. Guston and Daniel Sarewitz, "Real-Time Technology Assessment," *Technology in Society,* vol. 24 (2002), 93–109.

15. Edith Stokey and Richard Zeckhauser, *A Primer for Policy Analysis* (New York: W.W. Norton & Co., 1978), 5.

16. International Risk Governance Council, *White Paper on Nanotechnology Risk Governance,* Geneva, Switzerland: International Risk Governance Council, 2006, 53–56.

17. Maria C. Powell and Mathilde Colin, "Participatory Paradoxes: Facilitating Citizen Engagement in Science and Technology from the Top-Down?" *Bulletin of Science, Technology, and Society,* vol. 29, no. 4, August 2009, 325–342.

18. Maria C. Powell and Mathilde Colin, "Participatory Paradoxes: Facilitating Citizen Engagement in Science and Technology from the Top-Down?" *Bulletin of Science, Technology, and Society*, vol. 29, no. 4, August 2009, 329–330.

19. Maria C. Powell and Mathilde Colin, "Participatory Paradoxes: Facilitating Citizen Engagement in Science and Technology from the Top-Down?" *Bulletin of Science, Technology, and Society*, vol. 29, no. 4, August 2009, 340.

20. Richard Sclove, *Reinventing Technology Assessment: A 21st Century Model* (Washington, DC: Woodrow Wilson International Center for Scholars, April 2010), 6.

21. Richard Sclove, *Reinventing Technology Assessment: A 21st Century Model* (Washington, DC: Woodrow Wilson International Center for Scholars, April 2010), 8.

22. J. Clarence Davies, *Oversight of Next Generation Nanotechnology* (Washington, DC: Woodrow Wilson International Center for Scholars, April 2009), 26, 36.

23. J. Clarence Davies, *Oversight of Next Generation Nanotechnology* (Washington, DC: Woodrow Wilson International Center for Scholars, April 2009), 36.

24. National Nanotechnology Initiative, http://www.nano.gov/html/society/EHS.html.

25. Jeff Morris, "Global Engagement on Nano EHS: Role of the OECD in International Governance," keynote presentation to the annual meeting of the International Center for the Environmental Implications of Nanotechnology, University of California at Los Angeles, 12 May 2010.

26. Habermas, 57.

27. David Rejeski, "Managing the Molecular Economy," *Environmental Forum*, January/February 2010, Washington, DC: Environmental Law Institute, www.eli.org.

28. Daniel Sarewitz and Edward Woodhouse present a wide-ranging discussion of possible nanotechnology societal issues in "Small is Powerful," in Alan Lightman, Daniel Sarewitz, and Christina Desser, editors, *Living with the Genie: Essays on Technology and the Quest for Human Mastery* (Washington, DC: Island Press, 2003), 63–83.

29. Franz Geiger, "Nanotechnology in Climate Change—the Easy Way Out?" *Northwestern University Nanotechnology Town Hall Meeting*, video of presentation located at http://www.nsec.northwestern.edu/townhalls.htm.

Appendix

Charts Describing How Statements Can Be Organized

The following charts provide information regarding the risk statements analyzed in this work.

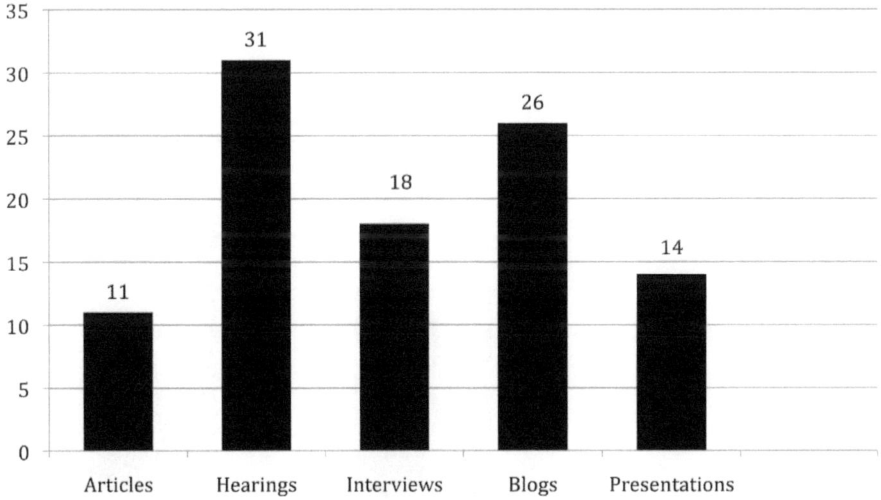

Figure 1A. Sources of Statements (as a percentage of total number of statements)

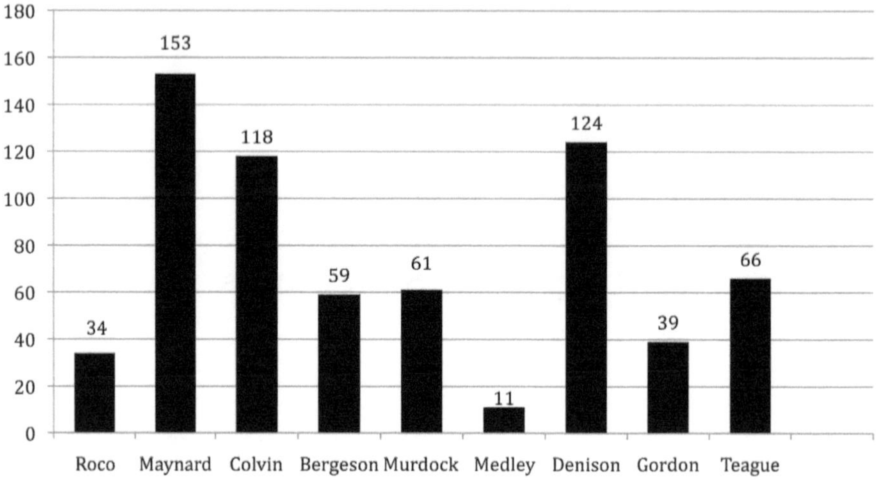

Figure 2A. Sources of Statements

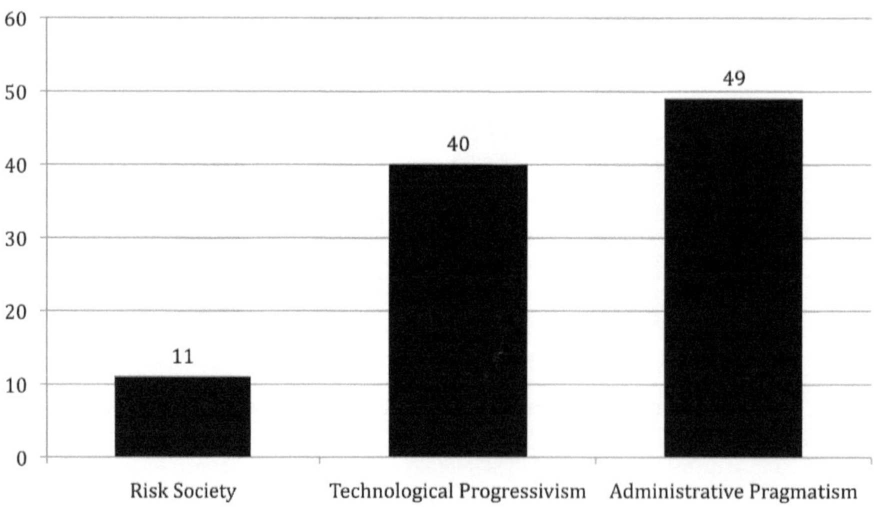

Figure 3A. Statement Within Discursive Categories (as percent of total number of statements)

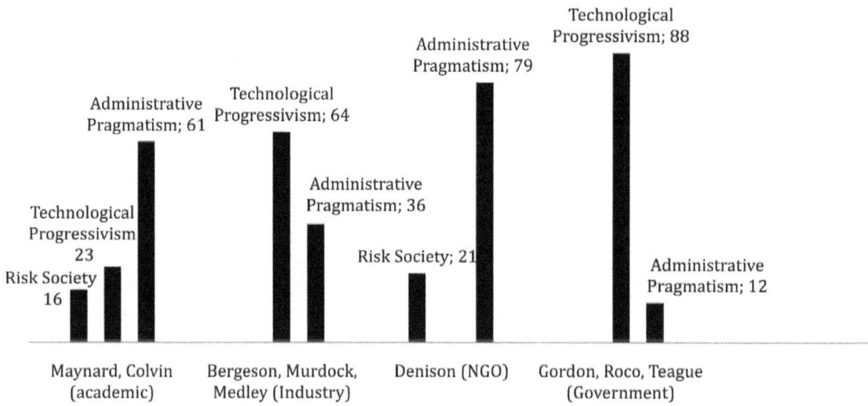

Figure 4A. Statements Within Discursive Categories, Grouped by "Sector" (as percent of that group's statements)

Bibliography

Agamben, Giorgio. *Homo Sacer: Sovereign Power and Bare Life*. Stanford: Stanford University Press, 1995.

Amato, Ivan. "Making the Right Stuff." *Science News*, vol. 136, no. 7, August 12, 1989, 108–110.

Anastas, P. T. and J. C. Warner, *Green Chemistry: Theory and Practice*, New York: Oxford University Press, 1998.

ASTI Nanotechnology Research Institute. "China Is To Become More Globally Competitive in Nanotech Industry." *Asia Pacific Nanotech Weekly*, vol. 33, article 33, 2005, http://www.nanoworld.jp/apnw/articles/library3/pdf/3-33.pdf.

Ball, Philip. "Synthetic Biology for Nanotechnology." *Nanotechnology* 16, R1-R8 (2005). Published online: 8 December 2004, http://iopscience.iop.org/0957-4484/16/1/R01.

Baxter, David. "Nanotechnology: An Insurer's Perspective." *Safenano.org*, February 2008, http://www.safenano.org/NanoInsurancePerspective.aspx.

Beck, Ulrich. *Ecological Enlightenment: Essays on the Politics of the Risk Society*. Atlantic Highlands, NJ: Humanities Press International, Inc., 1995.

Beck, Ulrich. *Risk Society: Towards a New Modernity*. London: Sage Publications, 1992.

Bergeson, Lynn. "Food and Drug Administration's Regulation of Nanotechnology." *Daily Environment Report* ISSN 1060-2976 BNA 9-22-08.

Bergeson, Lynn. "Nanotechnologies and FIFRA." *ChemAdvisory*, vol. 35, July 2006: 4. Originally published in the April 2006 issue of the Gradient Corporation EH&S Nano News.

Bloor, David. *Knowledge and Social Imagery, Second Edition*. Chicago: The University of Chicago Press, 1991.

Borradori, Giovanna. *Philosophy in a Time of Terror: Dialogues with Jürgen Habermas and Jacques Derrida*. Chicago: University of Chicago Press, 2003.

Bosso, Christopher J. *Pesticides and Politics: The Life Cycle of a Public Issue*. Pittsburgh: University of Pittsburgh Press, 1987.

Bosso, Christopher J., editor. *Governing Uncertainty: Environmental Regulation in the Age of Nanotechnology*. Washington, DC: RFF Press, 2010.

Bovenkerk, Bernice. "Is Smog Democratic? Environmental Justice in the Risk Society." *Melbourne Journal of Politics*, 2003. Full text available at http://www.thefreelibrary.com.

Brown, David. "Scientists Create Cell Based on Man-Made Genetic Instructions," *Washington Post*, 21 May 2010, www.washingtonpost.com.

Bullis, Kevin. "'Nano' Safety Recall: A product touted as 'nano' has hospitalized six German consumers, prompting more warnings over the dangers of nanomaterials." *MIT Technology Review*, online version, 10 April 2006, http://www.technologyreview.com/biomedicine/16681/1.

Burchell, Graham, Colin Gordon, and Peter Miller, editors. *The Foucault Effect: Studies in Governmentality*, Chicago: The University of Chicago Press, 1991.

California Institute of Science. *Engineering and Science*, vol. 23:5, February 1960.

Carson, Rachel. *Silent Spring*. New York, NY: Houghton Mifflin, 1962.

Chapman, Anne. *Democratizing Technology*. London: Earthscan, 2007.

Colvin, Vicki. "Regulation? Wait for Standardization, Commercialization." *The Environmental Forum®*. Reprinted by permission from the Environmental Law Institute. Washington, DC, March/April 2004.

Comp, Nathan J. "Harmful or not? That's hard to tell: Nanotech watcher Andrew Maynard assesses its risks." *The Daily Page*, 5 July 2007, http://www.thedailypage.com/daily/article.php?article=7792.

Crist, Eileen and H. Bruce Rinker, editors. *Gaia in Turmoil: Climate Change, Biodepletion, and Earth Ethics in an Age of Crisis*. Cambridge: The MIT Press, 2010 (Kindle version).

Davies, J. Clarence. *Oversight of Next Generation Nanotechnology*. Washington, DC: Woodrow Wilson International Center for Scholars, April 2009.

Denison, Richard. "A proposal to increase federal funding of nanotechnology risk research to at least $100 million annually." *Environmental Defense staff paper*. April 2005.

Desrosières, Alain. *The Politics of Large Numbers: A History of Statistical Reasoning*, translated by Camille Naish. Cambridge: Harvard University Press, 1998.

Douglas, Mary and Aaron Wildavsky. *Risk and Culture: An Essay on the Selection of Technical and Environmental Dangers*. Berkeley: University of California Press, 1982.

Dryzek, John S. *Discursive Democracy: Politics, Policy, and Political Science*. Cambridge: Cambridge University Press, 1990.

Dryzek, John S. *The Politics of the Earth: Environmental Discourses*. New York: Oxford University Press, 2005.

Duhem, Pierre. "Physical Theory and Experiment." In Martin Curd and J. A. Cover. *Philosophy of Science: The Central Issues*. New York: W. W. Norton and Company, 1998.

Durkheim, Émile. *The Elementary Forms of Religious Life* [1912]. Translation and introduction by Karen E. Fields. New York: The Free Press, 1995.

Elvin, George. "No nano in Magic Nano," NanotechBuzz, 27 May 2006, http://www.nanotechbuzz.com/50226711/no_nano_in_magic_nano.php.

Environmental Defense and DuPont. *Nano Risk Framework*. June 2007, www.nanoriskframework.com.

Environmental Working Group. "The Failed EPA Asbestos Ban." http://www.ewg.org/sites/asbestos/facts/fact5.php.

ETCGroup. www.etcgroup.org/upload/publication/pdf_file/610.

ETC Group. "Nanotech News in Living Colour: An Update on White Papers, Red Flags, Green Goo, Grey Goo (and Red Herrings)," *ETC Group Communiqué*, Issue # 85, May/June 2004, http://www.etcgroup.org/upload/publication/95/01/livingcolorfinal.pdf.

Executive Office of the President. National Science and Technology Council. Subcommittee on Nanoscale Science, Engineering, and Technology. *Strategy for Nanotechnology-Related Environmental, Health, and Safety Research*. February 2008, www.nano.gov/NNI_EHS_Research_Strategy.pdf.

Executive Office of the President, President's Council of Advisors on Science and Technology, *Report to the President and Congress on the Third Assessment of the National Nanotechnology Initiative*, 12 March 2010.

Ezrahi, Yaron. *The Descent of Icarus: Science and the Transformation of Contemporary Democracy*. Cambridge: Harvard University Press, 1990.

Federal Institute for Risk Assessment. "Nano particles were not the cause of health problems triggered by sealing sprays!" BfR press release, 26 May 2006, http://www.bfr.bund.de/cd/7842.

Foucault, Michel. *Discipline and Punish*. London: Tavistock, 1977.

Foucault, Michel. *History of Sexuality, Volume 1: An Introduction*. Translation by Robert Hurley. New York: Random House, 1978.

Geiger, Franz. "Nanotechnology in Climate Change—the Easy Way Out?" *Northwestern University Nanotechnology Town Hall Meeting*, video of presentation located at http://www.nsec.northwestern.edu/townhalls.htm.

Giddens, Anthony. *The Consequences of Modernity*. Stanford: Stanford University, 1990.

Gieryn, Thomas F. *Cultural Boundaries of Science: Credibility on the Line*. Chicago: The University of Chicago Press, 1999.

Gilbert, Natasha. "Crucial Data on REACH Not Disclosed." *Nature News*, Nature.com, 21 April 2010, http://www.nature.com/news/2010/100421/full/4641116a.html.

Goodman, Sara. "Sen. Lautenberg Introduces Chemicals Reform Bill, Saying Current Regulation 'Is Broken.'" *New York Times*, 15 April 2010, www.nytimes.com.

Graham, Sarah. "Nanotech: It's Not Easy Being Green." Interview with Dr. Vicki Colvin. *Scientific American*. www.scientificamerican.com. 28 July 2003.

Guston David H. and Daniel Sarewitz. "Real-Time Technology Assessment." *Technology in Society*, Volume 24 (2002).

Haack, Susan. *Defending Science—Within Reason: Between Scientism and Cynicism*. Amherst, NY: Prometheus Books, 2007.

Habermas, Jürgen. *The Postnational Constellation: Political Essays*. Translated by Max Pensky. Cambridge: The MIT Press, 2001.

Hackett, Edward J. et al., editors. *Handbook of Science and Technology Studies, Third Edition*. Cambridge: The MIT Press, 2007.

Hajer, Maarten A. *The Politics of Environmental Discourse: Ecological Modernization and the Policy Process*. Oxford: Oxford University Press, 1995.

Halfon, Saul. *The Cairo Consensus: Demographic Surveys, Women's Empowerment, and Regime Change in Population Policy*. Lanham, MD: Lexington Books, 2007.

Hall, Stuart. "Foucault: Power, Knowledge and Discourse." In Wetherell, Taylor and Yates.

Haraway, Donna J. *Simians, Cyborgs, and Women: The Reinvention of Nature*. New York: Routledge, 1991.

Harding, Sandra. *Whose Science? Whose Knowledge?* Ithaca: Cornell University Press, 1991.

Hardt, Michael and Antonio Negri. *Empire*. Cambridge: Harvard University Press, 2000.

Hart Research Associates. *Nanotechnology, Synthetic Biology, and Public Opinion: A Report Of Findings Based On A National Survey Among Adults, Conducted On Behalf Of: Project On Emerging Nanotechnologies The Woodrow Wilson International Center For Scholars*. Washington, DC: Hart Research Associates, 22 September 2009. http://www.nanotechproject.org/process/assets/files/8286/nano_synbio.pdf.

Hess, David J. *Science Studies: An Advanced Introduction*. New York: New York University Press, 1997.

International Risk Governance Council. *White Paper on Nanotechnology Risk Governance*. Geneva, Switzerland: International Risk Governance Council, 2006.

J. Craig Venter Institute, "First Self-Replicating, Synthetic Bacterial Cell Constructed by J. Craig Venter Institute Researchers," JCVI Press Release, 20 May 2010, http://www.jcvi.org/cms/press/press-releases/full-text/article/first-self-replicating-synthetic-bacterial-cell-constructed-by-j-craig-venter-institute-researcher/.

Jardine, Lisa. *Ingenious Pursuits: Building the Scientific Revolution*. New York: Nan A. Talese/Doubleday, 1999.

Jasanoff, Sheila. *The Fifth Branch: Science Advisors as Policymakers*. Cambridge: Harvard University Press, 1990.

Jasanoff, Sheila, editor. *States of Knowledge: The Co-Production of Science and Social Order*. London: Routledge, 2004.

Jasanoff, Sheila. *Designs on Nature: Science and Democracy in Europe and the United States*. Princeton: Princeton University Press, 2005.

Jaurand, Marie-Claude F., Annie Renier, and Julien Daubriac. "Mesothelioma: Do asbestos and carbon nanotubes pose the same health risk?" *Particle and Fibre Toxicology*, 2009, 6:16.

Kitcher, Philip. *Science, Truth, and Democracy*. Oxford: Oxford University Press, 2001.

Kleinman, Daniel Lee. *Science and Technology in Society: From Biotechnology to the Internet*. Malden, MA: Blackwell Publishing, 2005.

Kleinman, Daniel Lee. *Politics on the Endless Frontier: Postwar Research Policy in the United States*. Durham: Duke University Press, 1995.

Kroes, Peter and Anthonie Meijers, editors. "The Empirical Turn in the Philosophy of Technology." In Carl Mitcham, editor, *Research in Philosophy and Technology*, vol. 20, Amsterdam: Elsevier Science Ltd., 2000.

Kuzma, J. and Z. Meghani, "The Public Option: A possible change in the US risk-based decision making for emerging technological products: Compromised or enhanced objectivity?" *EMBO Reports* 10: 1–6 (2009).

Landmark Europe, "Nanotechnology in Consumer Products." July 2009, http://www.landmarkeurope.eu/nanotech.html.

Landy, Marc K., Mark J. Roberts, and Stephen R. Thomas. *The Environmental Protection Agency: Asking the Wrong Questions*. New York: Oxford University Press, 1990.

Latour, Bruno and Steve Woolgar. *Laboratory Life: the Social Construction of Scientific Facts*. Los Angeles, CA: Sage, 1979.

Latour, Bruno. *We Have Never Been Modern*. Translated by Catherine Porter. Cambridge: Harvard University Press, 1993.

Latour, Bruno. "Give me a laboratory and I will raise the world." *The Science Studies Reader*. M. Biagioli, editor, 258–275. New York: Routledge, 1999.

Laudan, Larry. "Demystifying Underdetermination." In Martin Curd and J. A. Cover. *Philosophy of Science: The Central Issues*. New York: W. W. Norton and Company, 1998.

Lewin, D.I. "DNA Computing." *Computing in Science & Engineering*. May/June 2002, vol. 4, issue 3, 5–8.

Lightman, Alan, Daniel Sarewitz, and Christina Desser, editors, *Living with the Genie: Essays on Technology and the Quest for Human Mastery*. Washington, DC: Island Press, 2003.

Litfin, Karen. *Ozone Discourses: Science and Politics in Global Environmental Cooperation*. New York: Columbia University Press, 1994.

Lockman, Paul R., Joanna M. Koziara, Russell J. Mumper, and David D. Allen. "Nanoparticle Surface Charges Alter Blood–Brain Barrier Integrity and Permeability." *Journal of Drug Targeting*, December 2004, vol. 12, no. 9–10: 635–641.

Lovy, Howard. "Being Mike Roco." *Howard Lovy's NanoBot*. 20 April 2004. http://nanobot.blogspot.com/2004/04/being-mike-roco.html.

Machamer, Peter and Michael Silberstein, editors. *The Blackwell Guide to the Philosophy of Science*. Malden, MA: Blackwell Publishers, Inc., 2002.

Mandel, H. George and Elliot S. Vesell. "From Progress to Regression: Biomedical Research Funding." *Journal of Clinical Investigation* 114(7) (2004): 872–876.

Marchant, Gary, Ann Meyer, and Megan Scanlon. "Integrating Social and Ethical Concerns Into Regulatory Decision-Making for Emerging Technologies." *Minnesota Journal of Law, Science, and Technology*, 2010:11(1): 345–363.

Marcuse, Herbert. *One-Dimensional Man*. Boston: Beacon Press, 1964.

Marx, Leo. *The Machine in the Garden: Technology and the Pastoral Ideal In America*. New York: Oxford University Press, 1964.

Maynard, Andrew. "Ten things everyone should know about nanotechnology safety." *Science 2020*. Blog posted at www.2020science.org.

Maynard, Andrew. "Tough Love for Science and Technology Innovation." *Science2020*. Blog posted on 10 December 2008, www.2020science.org.

Medley, Terry. "Nano Risk Framework—An EHS Assessment Tool." PowerPoint Presentation. Presented 20 March 2008, in Baltimore, MD, at a session titled, *Nanotechnology: The Future of EHS Regulatory Policy*.

Meghani, Z. and J. Kuzma. "Addressing the Regulatory Agency-Industry Revolving Door Problem for Emerging Technologies: Re-Conceptualizing Objectivity," *Journal of Agricultural and Environmental Ethics* 23(6): (2010).

Merchant, Carolyn. *The Death of Nature: Women, Ecology, and the Scientific Revolution*, San Francisco: Harper, 1989 [preface published 1990].

Merton, Robert K. *The Sociology of Science: Theoretical and Empirical Investigations*. Chicago: University of Chicago Press, 1973.

Miller, Peter and Nikolas Rose. *Governing the Present*. Malden, MA: Polity Press, 2008.

Mills, Sara. *Discourse*. New York: Routledge, 1997.

Minimum Information for Nanomaterial Characterization Initiative. "The Parameters List," http://characterizationmatters.org/parameters.

Morris, Jeff. "Global Engagement on Nano EHS: Role of the OECD in International Governance." Keynote presentation to the annual meeting of the International Center for the Environmental Implications of Nanotechnology. University of California at Los Angeles, May 12, 2010.

Nanotechproject.com.

Nanowerk.com. "Ten Things You Should Know About Nanotechnology," http://www.nanowerk.com/nanotechnology/ten_things_you_should_know_6.html.

National Council for Science and the Environment. CRS Report for Congress. "The Delaney Clause Effects on Pesticide Policy," http://www.cnie.org/NLE/CRSreports/Pesticides/pest-1.cfm.

National Nanotechnology Coordination Office. "Sen. Allen Announces Congressional Nanotechnology Caucus." NNCO press release, April 2004, http://www.nano.gov/html/news/releases/NNIConf04.html.

National Nanotechnology Initiative. "Capstone Meeting: Risk Management Methods & Ethical, Legal, and Societal Implications of Nanotechnology." 30–31 March 2010, http://www.nano.gov/html/about/symposia.html.

National Research Council. Committee on Toxicity. *Toxicity Testing in the 21st Century: A Vision and a Strategy*. Washington, DC, The National Academies Press, 2007.

National Research Council. *Risk Assessment in the Federal Government: Managing the Process*. Washington, DC: National Academies Press, 1983.

National Science, Engineering, and Technology Subcommittee. "Materials By Design: Report of the National Nanotechnology Workshop." June 11–13, 2003, http://www.nano.gov/NNI_Materials_by_Design.pdf.

Nel, Andre E. et al. "Understanding biophysicochemical interactions at the nano–bio interface." *Nature Materials* 8, 543–557 (2009). Published online: 14 June 2009, http://www.nature.com/nmat/journal/v8/n7/abs/nmat2442.html.

OECD. "Draft Guidance on Sample Preparation and Dosimetry for the Safety Testing of Manufactured Nanomaterials," OECD document ENV/CHEM/NANO(2009)7/REV2.

OECD. "Mutual Acceptance of Data (MAD)," http://www.oecd.org/document/41/0,3343,en_2649_34365_1890473_1_1_1_1,00.html.

OECD. "OECD Conference on Potential Environmental Benefits of Nanotechnology: Fostering Safe Innovation-Led Growth." 15–17 July, 2009, http://www.oecd.org/document/40/0,3343,en_2649_37417_42323688_1_1_1_1,00.html.

OECD. "Report Of An OECD Workshop On Exposure Assessment And Exposure Mitigation: Manufactured Nanomaterials." *Joint Meeting Of The Chemicals Committee And The Working Party On Chemicals, Pesticides And Biotechnology*, OECD document number ENV/JM/MONO(2009)18, 7 July 2009, http://www.oecd.org/dataoecd/15/25/43290538.pdf.

OECD. "Report of the OECD Workshop on the Safety of Manufactured Nanomaterials: Building Co-operation, Co-ordination and Communication." OECD document number ENV/JM/MONO(2006)19, 28 April 2006, http://www.olis.oecd.org/olis/2006doc.nsf/LinkTo/NT00000F92/$FILE/JT03208175.PDF.

OECD. "Safety of Manufactured Nanomaterials," http://www.oecd.org/topic/0,3373,en_2649_37015404_1_1_1_1_37465,00.html.

OECD. "Sponsorship Programme for the Testing of Manufactured Nanomaterials," http://www.oecd.org/dataoecd/53/41/43532644.pdf.

OECD. "Guidance Manual for the Testing of Nanomaterials: OECD's Sponsorship Programme; First Revision." 2 June 2010, http://www.oecd.org/officialdocuments/displaydocumentpdf/.

OECD. "OECD Programme on the Safety of Manufactured Nanomaterials 2009–2012: Operational Plans of the Projects." *Report to the Joint Meeting of the Chemicals Committee and the Working Party on Chemicals, Pesticides and Biotechnology.* OECD document ENV/JM/MONO(2010)11, 26 April 2010, http://www.oecd.org/officialdocuments/displaydocumentpdf?cote=ENV/JM/MONO%282010%2911&doclanguage=en.

Pitt, Joseph. "When Is an Image Not an Image?" *Techné* 8:3, Spring 2005, 24–33.

Poland, Craig A., et al. "Carbon nanotubes introduced into the abdominal cavity of mice show asbestos-like pathogenicity in a pilot study." *Nature Nanotechnology* 3, 423–428 (2008). Published online 20 May 2008. doi:10.1038/nnano.2008.111.

Popper, Karl. "The Problem of Induction." In Martin Curd and J.A. Cover. *Philosophy of Science: The Central Issues.* New York: W.W. Norton and Company, 1998.

Porter, Theodore M. *Trust in Numbers.* Princeton: Princeton University Press, 1995.

Powell, Maria C. and Mathilde Colin. "Participatory Paradoxes: Facilitating Citizen Engagement in Science and Technology from the Top-Down?" *Bulletin of Science, Technology, and Society,* vol. 29, no. 4, August 2009, 325–342.

Project on Emerging Nanotechnologies. "Nanotech-enabled Consumer Products Top the 1,000 Mark," http://www.nanotechproject.org/news/archive/8277/.

Project on Emerging Nanotechnologies, "Nanotechnology and Consumer Products," 18 August 2009, http://www.nanotechproject.org/process/assets/files/8278/pen_submission_cpsc.pdf.

Quine, W. V. "Two Dogmas of Empiricism." In Martin Curd and J. A. Cover. *Philosophy of Science: The Central Issues.* New York: W. W. Norton and Company, 1998.

Rae, Linda. "Nanotechnology in the Semiconductor Industry." Keithley Instruments, Inc., September 2006. www.keithley.com.

Rejeski, David. "Managing the Molecular Economy." *Environmental Forum,* January/February 2010, Washington, DC: Environmental Law Institute, www.eli.org.

Rose, Nikolas. "Governing by Numbers: Figuring Out Democracy." *Accounting, Organization, and Society,* vol. 16, no. 7, 673–692, 1991.

Rose, Nikolas. *The Politics of Life Itself: Biomedicine, Power, and Subjectivity in the Twenty-First Century.* Princeton: Princeton University Press, 2007.

Rosenberg, Alex. *Philosophy of Science: A Contemporary Introduction, Second Edition.* New York: Routledge, 2005.

Sarewitz, Daniel. *Frontiers of Illusion: Science, Technology, and the Politics of Progress*. Philadelphia: Temple University Press, 1996.

Sclove, Richard. *Reinventing Technology Assessment: A 21st Century Model*, Washington, DC: Woodrow Wilson International Center for Scholars, April 2010.

Scott, James. *Seeing Like a State: How Certain Schemes to Improve the Human Condition Have Failed*. New Haven: Yale University Press, 1998.

Shapin, Steven and Simon Schaffer. *Leviathan and the Air Pump: Hobbes, Boyle, and the Experimental Life*. Princeton: Princeton University Press, 1985.

Shapin, Steven. *The Scientific Life: A Moral History of a Late Modern Vocation*. Chicago: The University of Chicago Press, 2008.

Shrader-Frechette, Kristin. *Risk and Rationality: Philosophical Foundations for Populist Reforms*. Berkeley: University of California Press, 1991.

Silver Institute. Silver Nanotechnology Working Group. "Comments of the Silver Nanotechnology Working Group for Review by the FIFRA Scientific Advisory Panel; re: Re: FIFRA Scientific Advisory Panel; Notice of Public Meeting Docket ID: EPA–HQ–OPP–2009–0683," 15. www.regulations.gov and http://www.silverinstitute.org/images/stories/silver/PDF/epa_hq_opp2009.pdf.

Slovic, Paul. *The Perception of Risk*. London, UK: Earthscan, 2000.

Society of Manufacturing Engineers, http://www.sme.org/cgi-bin/get-event.pl?--001624-000007-019902--SME-.

Spradley, James P. *The Ethnographic Interview*. New York: Holt, Rinehart and Winston, 1979.

Stokey, Edith and Richard Zeckhauser. *A Primer for Policy Analysis*. New York: W.W. Norton & Co., 1978.

Stoll, Mark. *Protestantism, Capitalism, and Nature in America*. Albuquerque: University of New Mexico Press, 1997.

Stone, Vicki et al. *Final Report: Engineered Nanoparticles – Review of Health & Environmental Safety*. European Joint Research Center, August 2009, http://nmi.jrc.ec.europa.eu/project/ENRHES.htm.

Sunstein, Cass R. *Risk and Reason: Safety, Law, and the Environment*. Cambridge: Cambridge University Press, 2002.

Sunstein, Cass. *Going to Extremes: How Minds Unite and Divide*. Oxford: Oxford University Press, 2009.

University of Montana, video: "Creating 'Good' Technologies? An Emerging Technology in the 21st Century" with Vicki Colvin Debating Science University of Montana Center for Ethics 2008 Summer Series. http://video.google.com/videoplay?docid=-8198115874611299878#.

US Code of Federal Regulations, [15USC2603], [Page 1358-1362], Title 15, Commerce and Trade, Chapter 53, Toxic Substances and Control, Subchapter I, Control of Toxic Substances, Section 2603, Testing of chemical substances and mixtures, (a)(1)(A)(i), [1977],USGPO Online: http://frwebgate.access.gpo.gov.

US Congress. House Committee on Commerce. Subcommittee on Health and the Environment. *Food Quality Protection Act of 1995: Hearings before the Subcommittee on Health and Environment of the Committee on Commerce. House of Representatives*, 104th Congress, First session, on HR 1627, June 7 and June 29, 1995, Washington, DC: US Government Printing Office, 1995.

US Congress. House of Representatives. Committee on Science. "The Societal Implications of Nanotechnology." 108th Congress, First Session. 9 April 2003, Washington, DC: US GPO.

US Congress. House of Representatives. Committee on Science. "Environmental and Safety Impacts of Nanotechnology: What Research Is Needed?" 109th Congress, First Session. 17 November 2005, Washington, DC: US GPO.

US Congress. House of Representatives. House Committee on Science and Technology. Subcommittee on Research and Science Education. "Research On Environmental And Safety Impacts Of Nanotechnology: Current Status Of Planning And Implementation Under The National Nanotechnology Initiative." 110th Congress, First Session, 31 October 2007, Washington, DC: US Government Printing Office.

US Congress. House of Representatives. Committee on Science and Technology. "National Nanotechnology Initiative Amendments Act of 2008." 110th Congress, Second Session. 16 April 2008, Washington, DC: US GPO.

US Congress. Senate, Committee on Energy and Commerce. "Hearing on Developments in Synthetic Genomics and Implications for Health and Energy." 111th Congress, Second Session, 27 May 2010.

US Department of Health and Human Services. National Institutes of Health. *2008–2009 NIH Almanac*. NIH Publication no. 08–05, October 2008.

US Department of Health and Human Services. National Institutes of Health. National Cancer Institute. *Reducing Environmental Cancer Risk: What We Can Do Now*. President's Cancer Panel 2008–2009 Annual Report. Bethesda, MD: National Cancer Institute, April 2010.

US EPA. *Risk Assessment Principles and Practices*. EPA Document EPA/100/B-04/001, March 2004.

US EPA. "Risk Assessment Portal," www.epa.gov/risk.

US EPA. *A Summary of General Assessment Factors for Evaluating the Quality of Scientific and Technical Information*. EPA document number 100/B-03/001, *2003*. http://www.epa.gov/osa/spc/assess.htm.

US EPA. *Exposure Factors Handbook* (Final Report). US Environmental Protection Agency, Washington, DC, EPA/600/P-95/002F a-c, 1997.

US EPA. *Ecological Risk Assessment Guidelines*. 1998, http://cfpub.epa.gov/ncea/cfm/recordisplay.cfm?deid=12460.

US EPA. *Nanomaterial Research Strategy*. June 2009, EPA document number 620/K-09/011. The EPA strategy may be found at www.epa.gov/nanoscience.

US EPA. *Nanomaterial Research Strategy*. June 2009, www.epa.gov/nanoscience.

US EPA. "Endangerment and Cause or Contribute Findings for Greenhouse Gases under Section 202(a) of the Clean Air Act," http://www.epa.gov/climatechange/endangerment.html.

US EPA. "November 3–6, 2009: Assessment of Hazard and Exposure Associated with Nanosilver and other Nanometal Pesticide Products: A Consultation," http://www.epa.gov/scipoly/sap/meetings/2009/110309ameeting.html.

US EPA. "Risk Assessment for Toxic Air Pollutants: A Citizen's Guide." Originally published as EPA 450/3-90-024, March 1991, http://www.epa.gov/ttn/atw/3_90_024.html.

US EPA. *EPA Nanotechnology White Paper* (2007), www.epa.gov/osa.
US EPA. *Guidelines for Exposure Assessment*, 29 May 1992, Federal Register 57(104):22888-22938, EPA document number EPA/600/Z-92/001, www.epa.gov/risk/guidance.htm.
US EPA. *Nanoscale Materials Stewardship Program, Interim Report*, January 2009, http://www.epa.gov/oppt/nano/nmsp-interim-report-final.pdf.
US EPA. National Center for Environmental Assessment. www.epa.gov/risk.
US EPA. Nanotechnology grant solicitations, http://www.epa.gov/ncer/nano/solicitation/index.html.
US EPA. Office of Pesticide Programs, http://www.epa.gov/pesticides/factsheets/reg_fs.htm.
US EPA. Risk Assessment Forum. http://cfpub.epa.gov/ncea/raf.
US EPA. Risk Assessment Portal. http://www.epa.gov/risk/dose-response.htm.
US EPA. *Risk Characterization Handbook* (EPA 100-B-00-002, December 2000), http://www.epa.gov/osa/spc/pdfs/rchandbk.pdf.
US EPA. *Summary of General Assessment Factors for Evaluating the Quality of Scientific and Technical Information* (EPA 100/B-03/001, June 2003), http://www.epa.gov/osa/spc/pdfs/assess2.pdf.
US EPA. TSCA inventory http://www.epa.gov/oppt/newchems/pubs/invntory.htm.
US EPA, "Control of Nanoscale Materials Under the Toxic Substances Control Act." http://www.epa.gov/oppt/nano.
US EPA, "Proposed Decision Document for the Registration of HeiQ AGS-20, http://www.regulations.gov/search/Regs/home.html#documentDetail?R=0900006480b2f27e.
US EPA, "Nanomaterial Case Studies: Nanoscale Titanium Dioxide in Water Treatment and in Topical Sunscreen (External Review Draft)." EPA/600/R-09/057, July 2009, www.epa.gov/nanoscience.
US Food and Drug Administration. Center for Food Safety and Applied Nutrition. http://www.cfsan.fda.gov/~lrd/foodaddi.html.
US Geological Survey. "Toxic Substances and Hydrology Program: Five-Year Plan, 2007-2011." 15 March 2007,http://toxics.usgs.gov/pubs/pdfs/Toxics5-yrPlan2007-11_Final_3-15-07.pdf.
US Government Accountability Office. *Chemical Regulation: Options Exist to Improve EPA's Ability to Assess Health Risks and Manage Its Chemical Review Program*. GAO Report GAO-05-458. June 2005, http://www.gao.gov/new.items/d05458.pdf.
Weber, Max. *The Protestant Ethic and the Spirit of Capitalism*. Translation by Peter Baehr and Gordon C. Wells, New York: NY, Penguin Books, 2002.
Wetherell, Margaret, Stephanie Taylor, and Simeon J. Yates. *Discourse Theory and Practice: A Reader*. London, Thousand Oaks, New Delhi: Sage Publications, 2005.
Zaitchik, Alexander. "Russia Pours Billions in Oil Profits Into Nanotech Race." *Wired*, 1 November 2007, http://www.wired.com/science/discoveries/news/2007/11/russian_nano.

Index

actants, non-scientific, 140
actors, 8–9, 27n46, 51, 60, 75, 133, 140, 144, 158; mobilizing, 98; political, 42
administrative pragmatism, 13–14, 60–61, 76, 78–86, 107–8, 134–35, 138–39, 141, 151n47; blunting a technological progressivist position, 142; and capitalism, 107; and EPA, 133
administrative rationalism, 14
Agamben, Giorgio, 21, 92; *homo sacer*, 79
Amato, Ivan, 25, 177
American Chemistry Council, 11
American Physical Society, 4
Arizona State University, 25
artificial DNA, 2
asbestos, 57n78, 70, 109, 133, 170n9; ban, 50, 150n30; and mesothelioma, 50

Bacon, Francis, 98
Beck, Ulrich, 27n46, 33–34, 36–37, 49, 54, 80–81, 150n30, 156; reflexive modernization, 13, 33, 69, 139; risk society, 13, 33
benefit-risk assessment, 39–40, 140, 142
benefits, 5, 13, 22, 33, 39; distribution of, 3, 6, 13, 139, 150n37, 160; the myth of infinite, 72; nanomaterial, 45, 138; nanotechnology's, 13, 72, 85, 101, 128, 140; potential, 102; promised, 111; social, 13, 155
Bergeson, Lynn, 61, 76, 80–82
Bilirakis, Michael, 110
bioinformatics, 131
biopolitical power, 21, 92–93, 95–97, 105, 113, 119, 126, 129–30, 136, 154–55, 157, 168; considerations and implications, 99, 106, 129, 136; under the chemicals regime, 93
biopower. *See* biopolitical power
biotechnology, 26n23, 95
black boxing, 81, 88n57, 103, 122–24, 130
Bloor, David, 53
Bosso, Christopher J., 42, 132
boundaries, 14, 37, 42, 66, 84, 94, 96, 127, 131, 138, 158, 170n7; for STS scholarship, 22, 141, 158
Boyle, Robert, 98
Bridgman, Percy, 19

Brown, Wendy, 122
Bush, Vannevar, 78

calculated governance, 20–21, 40, 44, 51, 83, 86, 91–92, 99–100, 104–5, 111–13, 126, 137, 156,
170n9
calculating democracy. *See* calculated governance
cancer scares, 42, 93
Candidate Contaminant List (CCL), 40, 55n39
capitalism, 97–98, 104–7, 109, 111–13, 121, 157, 184, 186; global, 21, 86n7, 91–92, 99, 104–9, 111–13, 121, 125–26, 147, 156–57; hard shell of, 19, 29n64, 111
carbon nanotubes, 5, 46, 50, 57n78, 64, 109
Carson, Rachel, 93, 114n9
Chapman, Anne, riskiness, 37, 50, 52, 134
chemical regime, 93–95, 110, 112, 127, 129, 132, 136, 145, 148, 151n47, 167, 170n7
citizen juries, 162
Citizens Coalition on Nanotechnology, 162
Clean Air Act, 138
climate change, 125, 169
Colvin, Vicki, 63, 68, 72–74, 77, 100–101, 108–9, 114n13, 144, 147
Craig Venter Institute, 1, 24, 148
Crist, Eileen, 124–25

Danish consensus conferences, 163
Davies, J. Clarence, 163
debates. *See* risk debates
Delaney, James, 56n51
Denison, Richard, 63, 65, 68, 70, 79, 81, 101
Department of Agriculture, 9, 63
Desrosières, Alain, 115n41
discipline, 40, 42–44, 62, 81, 83, 112–13; behavioral, 42; legitimate biopolitical, 102; rhetorical, 42
discourse, 2, 13, 21, 27n37, 29n70; advancing, 2, 17, 21, 126, 154, 159–60, 168; defined, 60; limiting factors, 156; stability within, 137; tolerated, 123–24
discourse analysis, 17, 20–23, 59, 91, 95, 105, 132–33, 137, 140–41, 144, 153–55, 157–58
discursive categories, 13, 21, 61
dose-response assessments, 48, 143
Douglas, Mary, 34–36, 79; boundary actors, 81
Dryzek, John, 13, 27n39, 60, 86n7, 121, 128, 141, 149n2
DuPont, 10, 50, 63, 83
Durkheim, Émile, 18, 74

ecological effects, 15, 48; endpoints, 93; research, 48, 131
ecological modernization, 14, 180
economic considerations, 40, 95, 101; benefits, 35, 86, 100–101, 103, 108; competitiveness, 107, 109, 111
ED-DuPont Risk Framework, 10, 51, 83
effects, 41, 48, 156; environmental, 108; *See also* ecological effects; lung disease, 51, 109; neurological, 129; new toxicological, 138; potential human, 48; property-induced, 50; reproductive, 93
The Elementary Forms of Religious Life, 75
emerging technologies, risks of, 12, 168
engagement, 8, 144, 162; community, 166; cooperative, 67; direct, 102; public, 24, 162; structured citizen, 163
Enlightenment, 18, 21, 34, 76, 92, 112, 122, 124–25, 136, 149n2, 156
ENRHES report, 103
environmental debates, 2, 12, 35, 53, 126, 142, 159; policy debates, 2, 12, 141, 145, 161, 168; *See also* environmental discourse
Environmental Defense Fund (EDF), 10, 63; environmental discourse, 12,

14; changing, 23, 161; international, 86n7; western, 124
environmental policy, 3, 11, 22, 28n56, 69, 101, 127, 129, 140, 142, 146–47, 154, 158–59, 163; discourse on, 3, 11, 69, 105, 140, 147
Environmental Protection Agency (EPA), 3, 8–9, 11, 25, 31–32, 38, 40, 55n39, 73, 77–79, 81, 93, 103, 113, 122, 132–33, 164–65
environmental statutes, existing, 156
episteme, 27n37
epistemic possibility of harm, 36
European Commission, 7, 10, 103
expert systems, 14, 35, 43–45, 52, 81, 132, 156; risk analysis, 35; weak, 43, 94
exposure data, 32, 38–39, 59, 133
exposure science, 48–49, 103
Ezrahi, Yaron, 20, 41, 52, 80, 100–101, 104, 151n51

FDA, 9, 73, 79–81, 165
Federal Insecticide, Fungicide, and Rodenticide Act (FIFRA), 93, 97, 136
Feldman, Jay, 110
feminist writers, 122
Feynman, Richard, 4
Foucault, Michel, 19, 27n37, 92, 99, 126, 128, 137, 153, 155, 168
framings, 121, 126, 147, 161–62; dominant regulatory science/risk, 126; simultaneous, 13

Gaia theory, 123–26, 149n2, 178
Galison, Peter, 117n62
genetically modified (GM), 68, 70, 86n7, 95, 108, 114n13, 116n44, 136
geoengineering, 125
German Federal Institute for Risk Assessment, 1
Giddens, Anthony, 35, 43, 50, 132
Gieryn, Thomas, 18, 131; boundaries, 18, 131–32
Goldman, Lynn, 110
Gordon, Bart, 63, 142
governance, 19, 35, 37, 41, 94, 120, 132, 156, 158, 167–68; environmental, 53, 113, 156, 162
governance institutions, environmental, 159, 164
governmentality, 19, 27n46, 40, 44, 73, 75, 88n15, 99, 124, 130, 135, 156, 168
Gray, George, 110
green chemistry, 52–53, 94
Greenpeace, 10

Haack, Susan, 19, 140
Haas, Peter, 152n58
Habermas, Jürgen, 93, 167
Hajer, Maarten, 13, 128, 141; ecological modernization, 14, 53
Halfon, Saul, 29n69, 86n5, 128
Haraway, Donna, 64, 66, 97–98, 128
Hardt, Michael, 105, 115n41
Harvard Center for Risk Analysis, 110
hazards, 2, 36–38, 46, 57n78, 77, 133, 143, 154; environmental, 55n35; inherent, 68–69; new, 134; potential, 38
health effects, 36, 39, 42, 48, 55n35
health endpoints, 93
Hess, David, 18, 28n56
high-cost constructions, 104
homo faber, 16
homo sacer, 79
Hume, David, 116n52

imaging devices, 49–50
industrial chemicals, 2, 20, 35, 37, 48, 59, 63, 93, 132, 159; managing, 136; regulating, 94
influential participants, 22, 145
Integrated Risk Information System (IRIS), 31
interests, 14, 81, 123; alignment with actors, 144, 160; business, 138, 140; vested, 159
iron cage, capitalism's, 24, 157; *See also* capitalism, hard shell of

Jardine, 117n62
Jasanoff, Sheila, 19, 36–37, 40–41, 43, 94, 128–29, 131

Kitcher, Philip, 140
Kleinman, Daniel, 13, 15, 71, 78, 84, 95, 128; scientism and technological progressivism, 131
Kroes, Peter, 4
Kuzma, Jennifer, 25, 170n7, 181

Landy, Marc, 133, 150n30, 181
Latour, Bruno, 64, 66, 97, 122, 128, 140
Lautenberg, Frank, 94
Litfin, Karen, 28n54, 96, 124–25, 128, 144, 147, 158

Magic Nano, 1
Marcuse, Herbert, 105, 111, 115n41
material characterization, 46–47, 49, 53
material properties, 4, 50, 97, 109, 134
Maynard, Andrew, 63, 65, 68–70, 82, 84, 107, 142, 144
Medley, Terry, 63
Meijers, Anthonie, 25, 180
Merchant, Carolyn, 98–99, 122, 128
Merton, Robert, 75–76
metal catalysts, 5
metal oxides, 5, 103
metals, 5, 46
metaphors, 21, 49, 61–62, 68, 76–77, 82, 157
microscopy, 13, 47; confocal, 46; electron, 4–5, 46, 52
Miller, Peter, 29n65, 40, 73
Mills, Sara, 60, 128, 137
models, 31, 44, 47, 49, 67, 125
modernity, advanced, 34
Muir, John, 122
Murdock, Sean, 63, 83, 108

NanoBusiness Alliance, 11, 63, 108
Nano Cafes, 162–63
nanomaterials, 5, 35, 46; benign, 77; engineered, 65, 69, 83; new, 73, 160; silver, 97; *See also* nanoparticles
nanomaterials research, 108
nanoparticles, 5, 12; detecting airborne, 49; engineered or manufactured, 15, 67, 134, 142; first-generation, 7; functionalized, 7; incidental anthropogenic, 134; industry patents, 68; pictures of, 49, 52; properties of, 8, 50, 63, 146; transformation of, 47; *See also* nanoscale particles
nanoscale mirrors, 125, 169
nanoscale particles, 5, 45, 134, 137; non-engineered, 134
nanotechnology, 4; debates, 11, 16, 53, 60–61, 64, 92, 96, 130, 136, 147, 157; discourse, 96, 143; economic benefits of, 142, 157; environmental implications of, 1, 123, 126, 164; governance, 164–65; grants, 143; industry, 63, 72; knowledge brokers, 146; language about, 20; patents, 5; policy, 17, 130, 138; products, 7, 15, 51, 81, 113, 117n60; research, 5, 9, 102, 107; risk language, 16, 64, 67, 83, 91, 96, 128, 134, 139, 141, 157–59; risks, 50, 52
Nanotechnology Environmental and Health Implications working group, 9, 165
nano-TiO2, 103–4
National Academy of Sciences, 11, 38
National Institute for Environmental Health Sciences, 9, 11
National Institute for Occupational Safety and Health, 9, 11, 63
National Institute for Standards and Technology, 5, 63
National Institutes of Health, 78
National Nanotechnology Coordination Office (NNCO), 24, 63
National Science Foundation, 9, 11, 25, 63, 78, 165
Negri, Antonio, 105
non-governmental organizations (NGOs), 7, 9, 11, 25, 35, 81, 93, 100–102, 104, 155, 165, 167;

environmental, 8, 10, 15, 50–51, 81, 85, 101, 130, 138, 156

Obama, Barack 123
Occupational Safety and Health Administration (OSHA), 9, 73, 165
Office of Chemical Safety and Pollution Prevention, 93
Office of Research and Development, 7, 93
Organization for Economic Cooperation and Development (OECD), 7–11, 23, 26, 56–57n78, 63, 93-94, 112, 114n13, 167, 182–83; mutual acceptance of data, 94; sponsorship program, 56; test guidelines, 9; working party on manufactured manomaterials, 63, 165
outformations, 102–3, 113, 151n51
overlap statements, 141–42, 158

Pasteur, Louis, 140
PEN database, 25n16
photocatalytic properties, 103
Pitt, Joseph, 49
policy, 4; federal, 3; government energy, 92; industrial, 105; nanotechnology risk, 133; regulatory, 40, 139; science-based, 84–85; targeted, 127, 153
policy discourse, 17, 27n45, 119, 125, 142, 144; advancing nanotechnology, 167; dominant public, 52; enhancing environmental, 120, 147, 162
policy initiatives, 22–23, 159
political order, 99; constructing, 104
politics, technicization of, 42
politics of life, 21, 100
Popper, Karl, 116n52
Porter, Theodore, 19, 42, 44, 80, 99, 102, 132
Potter, Jonathan, 16, 141
Powell, Maria C., 162
power, 14; bio-political, 130; calculations of, 21, 92; global capitalism, 106, 108; micro physics of, 159, 168
practices: discursive, 91; knowledge-generation, 94, 155; long-standing scientific, 129; reflexive, 160
precautionary principle, 108
President's Council of Advisors on Science and Technology, 107
probabilities, calculating, 38, 42
progress: human, 62, 74; societal, 85; technological, 62, 71–72, 74
progressivism, 62
Project on Emerging Nanotechnologies, 63
pTA, 163

quantification, 40, 44, 61, 97, 99, 133; metrics, 82
quantum effects, 49–50, 67, 97, 102
quasi-governmental bodies, 156

rationality, one-dimensional, 111
reflexivity, 28n56, 71, 128, 159; *See also* Beck's reflexive modernization
regulation, 2, 31, 43–44; enhanced, 33; environmental, 96
regulatory science, 52, 130; assessments, 2; data, 21–22, 59, 81–82, 84, 109, 111, 133, 135–36, 143, 159, 167; disciplines, 48; endpoints, 129, 133; expert system, 51
regulatory science data: collection, 93; new, 136; using, 137
Rejeski, David, 168
representation, collective, 76, 100
representations, 42, 74; calculated, 132; shared cultural, 19
représentations collectives, 18
research, 101; academic, 11, 165; basic, 11; ethnographic, 145
Rice University, 63
risk analysis, 36, 43–44, 59, 106; Harvard Center for, 110
risk assessment, 1–2, 7, 20, 35, 38–41, 44, 46, 50–55n39, 93, 95, 110; anticipatory, 70; chemical, 52; environmental, 38;

federal, 43, 55n36, 93, 154, 166; for nanomaterials, 50–51, 53; quantitative, 34, 51–52, 100, 120, 170n9; risk characterization, 120; *See also* risk analysis
risk avoidance, 74, 81
risk debates, 3, 15, 21, 36, 53, 62, 81–82, 91, 110, 119, 132, 134, 142, 145, 153; broadening, 3, 120–21, 126, 133, 155–56, 164; *See also* risk discourse
risk discourse, 2, 17, 20, 23, 37, 53, 86n7, 94, 104, 120, 137–40, 145–46, 153, 155, 167
risk information, 14, 23, 43, 77, 81, 102, 132, 135, 143, 159, 167
risk management approaches, 7, 50
risks, 67; characterization of, 39–40, 94, 130; chemical, 35, 38, 41, 51, 94, 133; defining, 96, 128; nanotechnology, 15, 22–23, 69, 101, 119, 127, 132, 136
risk society, 13, 16, 33, 60, 65, 71, 80; discursive category language, 13, 16, 65–69, 72, 84–85, 107, 128, 139–40, 142
Roco, Mihail, 63, 72

Safe Chemicals Act, 94
safety, 2, 8; nanomaterial, 51, 73; technology's, 135
safety finding, 44
Sarewitz, Daniel, 15, 72, 78, 109, 171n28
Schaffer, Simon, 29n60, 75
scholarship, 4, 13; applying STS, 147; existing STS, 61; policy entrepreneur, 144
Science and Technology Studies (STS), 12, 23, 25, 157–58, 166
science's coextensive development with capitalism, 98
scientism, 13, 15, 22; discourse, 125; and regulatory science, 131; story line, 60, 81, 91, 96, 139; strength of, 130, 158
Sclove, Richard, 163
Scott, James, 19, 41, 82
Shapin, Steven, 75, 143, 147
Shrader-Frechette, Kristin, 35, 37
Spradley, James P., 145
Stickle, Warren, 110
Stokey, Edith, 161
Sunstein, Cass, 138
synthetic biology, 2, 23, 139, 148, 168, 170n9, 180

Teague, Clayton, 63, 73, 78, 111
Toxic Substances Control Act, 31-32, 93, 136; reform of, 94–95
Tyndall, John, 131

United Nations, 167
United States Geological Survey, 40
United States Bureau of Labor Statistics, 100
United States chemicals (TSCA) inventory, 31, 150n27

value of information methodologies, 101
venture capital funding, 107

Waxman, Henry, 2
Weber, Max, 19, 29n65, 104, 111, 186
Wetherell, Margaret, 16, 128, 141, 153
White House nanotechnology policy group, 79
Wildavsky, Aaron, 34–36, 79
Woodrow Wilson International Center For Scholars, 63
Woolgar, Steve, 88n57, 122
Working Party on Manufactured Nanomaterials (WPMN), 8–9, 11, 93, 167
wow-to-yuk, 69

About the Author

Jeffery T. Morris is a science and technology studies (STS) scholar and environmental protection professional. He has worked at the US Environmental Protection Agency in a number of capacities, including as national program director for nanotechnology research. Dr. Morris received his PhD from Virginia Tech, and also has graduate and undergraduate degrees from the University of Maryland, Ohio University, and Western Washington University. He and his wife Tracie live in Arlington, Virginia. The views in this book are his own and do not necessarily reflect the perspectives or policies of EPA or the US government.

www.ingramcontent.com/pod-product-compliance
Ingram Content Group UK Ltd.
Pitfield, Milton Keynes, MK11 3LW, UK
UKHW041629031026
15587UKWH00001B/159